BOMBS AND BOMBINGS

Fourth Edition

BRODIE'S

BOMBS AND BOMBINGS

A Handbook to Protection, Security, Detection, Disposal and Investigation for Industry, Police and Fire Departments

By

COLONEL JIM SMITH, MSS, DABCHS, CLEE, CPC

Public Safety Director
Cottonwood, Alabama

CHARLES C THOMAS • PUBLISHER, LTD.
Springfield • Illinois • U.S.A.

Published and Distributed Throughout the World by

CHARLES C THOMAS • PUBLISHER, LTD.
2600 South First Street
Springfield, Illinois 62704

ISBN 978-0-398-09093-7 (paper)
ISBN 978-0-398-09094-4 (eBook)

First Edition, 1972
Second Edition, 1996
Third Edition, 2005
Fourth Edition, 2016

Library of Congress Catalog Card Number: 2015028934

With THOMAS BOOKS *careful attention is given to all details of manufacturing and design. It is the Publisher's desire to present books that are satisfactory as to their physical qualities and artistic possibilities and appropriate for their particular use.* THOMAS BOOKS *will be true to those laws of quality that assure a good name and good will.*

Printed in the United States of America
CR-R-3

Library of Congress Cataloging-in-Publication Data

Smith, Jim (Public safety director)
Brodie's bombs and bombings : a handbook to protection, security, detection, disposal and investigation for industry, police and fire departments / by COLONEL Jim Smith, MSS, DABCHS, CLEE, CPC, Public Safety Director, Cottonwood, Alabama. -- Fourth Edition.
pages cm
Revised edition of Bombs and bombings : a handbook to protection, security, disposal, and investigation for industry, police and fire departments / by Thomas Graham Brodie. c2005
Includes index.
ISBN 978-0-398-09093-7 (pbk.) -- ISBN 978-0-398-09094-4 (ebook) 1. Bombs. 2. Explosives. 3. Bomb reconnaissance. 4. Bombings. I. Brodie, Thomas G. Bombs and bombings. II. Title. III. Title: Bombs and bombings.
TP270.B76 2015
623.4'51--dc23

2015028934

PREFACE

When asked by Michael Thomas of Charles C Thomas Publisher to update and revise *Bombs and Bombings* by Captain Thomas Brodie, I was pleased to have an opportunity to assist in the process. I had the pleasure of meeting and interacting with Captain Brodie twice during my career. I met him during presentation of research while lecturing at two IABTI conferences. Captain Brodie was the epitome of a professional.

He asked incisive questions regarding the blast and fragment mitigation methods I had researched and tested. He made several suggestions, which were helpful. The research was published internationally and resulted in the issuance of a patent for the process.

I find it an honor to follow in the footsteps of a "giant" in the law enforcement bomb squad community. His work is excellent but following his death had not been updated in several years. My mission was to bring the textbook into the current era of bomb squad operations without compromising operational security. The material presented is open source materials and does not address sensitive law enforcement methods. I am shocked at the volume of sensitive material available on the Internet. If in doubt, perform a search with the term "pipe bomb" or "making TATP" and you will be amazed at the information available, albeit some, which if followed, would result in injury or death.

Captain Brodie was central in the modernization of the law enforcement bomb squad. He contributed ideas he had acquired through practical applications. His ideas were the forerunners of today's bomb squad. However, he paid a serious price for his service. Captain Brodie was injured on several occasions during his career while investigating more than 350 bombings and disposing of more than 4,000 IEDs, along with tons of explosives. His methods following his injuries to prevent such along with his sharing of experiences have no doubt saved the lives of bomb technicians and law enforcement personnel.

Today, bomb technicians stand on the shoulders of such men as Thomas Brodie. His work led to his award as a Knight of the British Empire for his role in protecting British interests during the Cuban terrorism of the 1960s. He has been called a "hero" by the media and such a description is not unwarranted.

I have worked to revise and update this text to honor Captain Thomas Graham Brodie.

J.S.

ACKNOWLEDGMENTS

I have to thank the members of the Dothan Alabama Bomb Squad for their valuable assistance. Lt. Scot Heath, bomb squad commander; Lt. Brian Smith, assistant bomb squad commander; Cpl. Joseph Evans, bomb technician; retired Officer Glen Ketchum, CBRNE specialist; Officer Jeremy Conner; Officer Jeremy Wallace; retired Cpl. John Skipper, bomb technician and IED constructionist extraordinaire; retired post blast investigator Sergeant Gary Coleman; retired Bomb Assessment Officer Jason Youngblood; and retired Bomb Assessment Officer Kevin McKee. I would also like to thank the US Navy EOD personnel from Kings Bay Georgia and Army EOD personnel from Fort Benning Georgia for their input. Unfortunately for security reasons I cannot name them.

CONTENTS

Page

ILLUSTRATIONS

BOMBS AND BOMBINGS

Chapter 1

HISTORY OF BOMBS AND BOMBINGS

The first explosive described is black powder. This mixture was known by a variety of names. Black powder, or as it became to be known later as "gunpowder," is a mixture of sulfur, charcoal, and potassium nitrate. Black powder is extremely sensitive to friction, spark, flame, or heat. Chinese historical documents chronicle several events in which black powder storage magazines exploded with catastrophic results. These are early experiences with the instability of explosives. Gunpowder is mentioned in some Chinese texts as early as the mid-400s BCE. Some sources cite the earliest bombings as military actions in China during the 1200s as bamboo tubes filled with gunpowder. During this period, cast iron containers filled with gunpowder are believed to have been introduced and used in naval warfare in addition to land warfare.

During the 1400s, cannons using gunpowder were coming into common use. Between this time and the 1700s, firearms using gunpowder proliferated and hand grenades using gunpowder were introduced. Evidence exists that pirates used hand grenades in their attacks against merchant shipping during the 1600s and 1700s. During the early 1600s, gunpowder began to be used in mining as a blasting tool. Rockets appeared in the late 1700s and were used by the British in the form of Congreve Rockets against Fort McHenry as noted in America's National Anthem. Paper cartridges for firearms were developed during the 1700s. In 1846, nitrated cotton, called gun cotton, was developed which was smokeless and at the same time, nitroglycerin was developed. Nitroglycerin-based dynamite was developed. Smokeless gunpowder, the forerunner of modern smokeless gunpowder, is developed in the 1870s. Ammonium nitrate becomes a common component in dynamite by the 1880s. Binary explosives are used in the 1880s. Detonating cord is in use by 1902. By 1914, Germany had developed trinitrotoluene (TNT). Pentaerythritol tetranitrate (PETN) detonating cord is developed in 1938. Other explosives followed in WWII with many plastic explosives developed during this time. Prilled ammonium nitrate (AN) begins to replace dynamite in the 1950s for mining purposes. In 1969, emulsion explosives were marketed for mining. In the 1970s, black powder is supplanted by compounds adding graphite and potassium perchlorate to reduce sensitivity and improve its performance in firearms. Pyrodex® is an example of one of the several black powder substitutes available.

The United States has a long history of bombings and unintentional explosions as do many locales. Some of the more substantial and interesting bombings are discussed below with data drawn from open sources.

Early Bombings

1881–A hand-delivered bomb was used by a left-wing terrorist to assassinate the tsar of Russia. This is an early use of a person-borne, improvised explosive device (PBIED) as the assassin hand delivered the device and was killed in the ensuing explosion. Three bombers were present to assure the murder of the tsar. The tsar is killed after he exits his bombproof carriage.

May 4, 1886–A bomb is thrown at police during a labor rally in Chicago. During the event, seven police officers are killed and 11 others killed with more than 100 injured. The device is an improvised explosive device filled with dynamite encased in a brittle iron container hand delivered.

October 1, 1910–Union members angry at The Los Angeles Times for its supposed antiunion articles planted 16 sticks of dynamite in an alley adjacent to the building. The explosion killed 20 and injured nearly 100 others. Extensive damage to the building was done by the blast.

July 2, 1915–A German professor, Frank Holt, tried to stop American support of the Allies in WWI by detonating a bomb in the US Senate Reception Room. He used a sulfuric acid delay with high explosive (likely dynamite) to create the bomb. No injuries resulted.

July 22, 1916–The Preparedness Day parade was designed to prepare those living San Francisco for entry of the United States into World War I. Antiwar activists detonated a bomb killing 10 and injuring approximately 40. The bomb was either delivered as an emplaced improvised explosive device (IED) consisting of dynamite in a steel pipe or was alternatively dropped into the parade from a roof top, as accounts vary.

November 24, 1917–A bomb explodes in a Milwaukee police station, killing nine officers and a civilian. Anarchists were suspected. The device was found at a church and was delivered to the police department by the church's janitor. It exploded as officers were examining the package. The bomb was likely a pipe bomb filled with gunpowder.

1919 Anarchists bombings–These occurred during April through June of 1919. The anarchists were targeting important government and business officials. Roughly 30 bombs were mailed during the first attacks. The devices are described as a wooden block with dynamite contained within with an acid initiator activated as the package was opened allowing the acid to discharge blasting caps. Two injuries resulted and the post office was able to impound most of the devices. The June 2, 1919 attacks of the residences of several government officials resulted in minor injuries and the death of one of the bombers when the much larger and more powerful device detonated while in the bomber's possession. One residence was largely demolished. These bombs were hand delivered.

September 16, 1919–A bomb exploded in New York City's Wall Street area, killing 40 and injuring almost 300 people. The bombing is a horse-drawn carriage with an estimated 100 pounds of dynamite and a substantial amount of iron to serve as fragments. Anarchists are suspected in the bombing. This is an early use of a vehicle-borne, improvised explosive device (VBIED).

May 18, 1927–The Bath School is bombed. The school treasurer placed hundreds of pounds of dynamite in the school and detonated it. He drove back to the scene with his car containing a large amount of dynamite and detonated it killing himself and the principal of the school. This blast killed a student who had survived the initial blasts. Reports place the number of deaths at 45 with 38 listed as students. The dispute resulted from property tax issues. This is an early use of a VBIED.

October 10, 1933–A commercial Boeing 247 is destroyed in midflight over Indiana by a nitroglycerin bomb. The seven people aboard are killed. This is alleged to be the first proven case of air sabotage in the history of aviation in the US. The identity of the perpetrator and the rationale for the attack are unknown. The FBI believed nitroglycerin was the explosive used and the device was placed in the baggage compartment or toilet of the aircraft.

July 4, 1940–Two NYPD bomb squad officers are killed and other officers critically injured while attempting a hand render safe procedure on a bomb at the British Pavilion at the World's Fair.

1940 through 1956

George Metesky, also known as the "Mad Bomber," a former Consolidated Edison employee who had been injured and believed he had not been compensated properly, used bombs to attract attention. Metesky attacked New York City landmarks. He is alleged to have delivered 33 bombs with 22 detonating, injuring 15. He was caught and sentenced to a mental health facility. His bombs were pipe bombs filled with gunpowder using a mechanical timer. Metesky wrote numerous letters to newspapers and did provide some warning regarding impending detonations.

1950 through 1975

The FALN Bombing Campaign was one of longer running terrorist attacks in US history. These attacks included an attempt in 1950 on the life of President Truman, and in 1954, an attack on the House of Representatives in which five congressmen were injured. The FALN took credit for the bombing of a restaurant in New York City in which four were killed and roughly 50 injured. In Chicago on June 14, 1975, before the Puerto Rican Independence Day Parade, bombs were detonated in nearby buildings injuring four persons. In October of 1975, several other buildings were targeted in Chicago. These bombs used dynamite. On the same day, five bombings occurred in New York City and two bombs were detonated in Washington, DC with no injuries.

November 1, 1955–United Airlines Flight 629 was destroyed by an individual who had problems with his mother. He insured his mother with several policies, sent her on a trip with a present. The present was a bomb which exploded destroying the aircraft and killing the 44 persons aboard. Dynamite was the suspected explosive.

September 15, 1959 – Poe Elementary School bombing leaves six dead 19 injured, when a parent detonated an explosive device on the school playground because his son was denied entry to the school. The suitcase containing the device was believed to have used six sticks of dynamite. This is an example of a PBIED.

1960s through 1970s

May 22, 1962–Continental Airlines Flight 11 was destroyed by a suicide bomber trying to accrue money for his wife and children. The device as believed to contain six sticks of dynamite. The explosion killed 45 people. This is an example of a PBIED.

1960–The Sunday Bomber sets off a series of bombs in New York City subways and on ferries killing one and injuring 51. He uses pipe bombs.

August 7, 1969–Some 20 persons were injured in the Marine Midland Building bombing in New York City. An estimated 25 sticks of dynamite were used in the bomb. The attack is by leftist radicals the Weather Underground.

August 8, 1969–The US Department of Commerce building in New York City is damaged by a bombing. Leftist radicals from the Weather Underground are suspected in the attack.

September 18, 1969–The leftist radicals from the Weather Underground attack the federal building in New York City.

October 7, 1969–The armed forces induction center is attacked with a bomb by leftist radicals from the Weather Underground.

November 12, 1969–The Manhattan Criminal Court is attacked with a bomb by leftist radicals from the Weather Underground.

March 6, 1970–The two radicals from The Weather Underground kill themselves by unintentionally detonating a bomb they are assembling in a building in New York City.

August 24, 1970–University of Wisconsin in Madison is attacked with a VBIED in protest against the Vietnam War. The blast kills a professor and injures several others. A stolen van with an estimated

2,000 pounds of ammonium nitrate fuel oil mixture (ANFO) is used in the attack. The attackers are antiwar protesters. As part of the attack, they stole a light aircraft and dropped IEDs on an US Army Ammunition plant, but the devices fail to explode.

Summer 1974–The Alphabet Bomber bombs the Pan Am Terminal at Los Angeles on August 6, 1974 killing three and injuring 36. He fire bombs the houses of a judge and two police commissioners and burned down two apartment buildings. He also threatened to attack with nerve agent. One bomb was rendered safe by the LAPD which characterized the bomb as the largest seen at the time by the LAPD bomb squad. The bomber was trying to change immigration law.

January 24, 1975–The historic Fraunces Tavern in New York City is bombed by the FALN, killing four. The device is alleged to be 10 pounds of dynamite left in an attaché case in the entry of the business. This is one of 49 bombings attributed to the FALN in New York City between 1974 and 1977 with a total of 130 bombings nationwide.

December 29, 1975–A bomb hidden in a locker exploded at the TWA terminal at New York's LaGuardia Airport. The device suspected of containing upwards of 25 sticks of dynamite kills 11 and injures 75. The metal fragments from the locker and glass fragments account for many of the injuries and deaths. No arrests are made.

May 25, 1978–UNABOMBER sends his first mail bomb to Northwestern University. These early devices are small pipe bombs which are hand crafted. The bombs use metallic pipe with wooden end caps. This lack of a strong end cap reduces the explosive power of the bomb. The ignition system is a nail attached to a strong rubber band which was released when the package was opened. The nail used friction to ignite matches contained within the device to detonate it. As UNABOMBER progressed, he began to use batteries to heat a wire filament to detonate the device. The first device injures a police officer investigating the suspicious package.

May 9, 1979–UNABOMBER's second mail bomb injures a graduate student at Northwestern University.

November 15, 1979–A bomb on board a flight explodes causing several passengers to be treated for smoke inhalation. This is the first attack by UNABOMBER of an airliner.

1980s through 1990s

June 3, 1980–A bomb detonated in the Statue of Liberty's Story room causing damage. No one is injured as the device detonated after closing. No arrests are made.

June 10, 1980–UNABOMBER sends a mail bomb to the president of an airline. The device injures the individual but does not kill him.

May 16, 1981–Kennedy Airport New York City, a bomb is detonated a in the restroom in the Pan Am terminal. The bomb kills one. A group supporting independence for Puerto Rico claims responsibility. A second device is found near a boarding gate but is removed prior to detonation. No arrests were made.

October 8, 1981–The UNABOMBER's letter bomb to the University of Utah is detected. The device is rendered safe.

May 5, 1982–A UNABOMBER mail bomb is sent to Vanderbilt University. The blast injures a university secretary.

July 2, 1982–The UNABOMBER sends a letter bomb to University of California Berkeley where it injures a professor.

April 18, 1983–The Beirut Embassy in Lebanon is destroyed by a suicide VBIED. Islamic Jihad claims responsibility for the bombing. Some 63 are killed, including 17 Americans. A vehicle packed with approximately 2000 pounds of explosives is used. The pattern of the bombing mirrors that of the Murrah Building bombing as the vehicle is parked under the portico of the structure.

October 23, 1983–Shiite suicide bombers in Beirut Lebanon drive a truck into the US military barracks and detonate a bomb. Contemporaneously, a second bomb targets French paratroopers at another location. Some 241 Marines are killed and 58 French paratroopers are killed while more than another 100 Marines are injured. The explosion is estimated to have the force of roughly 20,000 pounds of TNT. The device is PETN explosive enhanced with butane containers sitting on concrete slabs placed in a stake bed truck. The truck drove through a concertina wire barrier and a guard shack to enter the lobby of the building where it detonated.

November 7, 1983–A bomb blows a hole in a wall outside the Senate chamber at the Capitol in Washington. No one is hurt. Two leftist radicals plead guilty.

December 12, 1983–The US Embassy in Kuwait City, Kuwait is attacked by Shiite suicide truck bombers. The attacks result in five deaths and around 80 injured.

September 20, 1984–In Beirut, Lebanon, a truck bomb exploded near the US Embassy Annex killing 24 persons. Reports vary as to the type and quantity of explosives with some reports citing up to 1,000 pounds of explosives while others cite lesser amounts.

May 15, 1985–UNABOMBER mailed a bomb to the University of California at Berkeley. It detonates and seriously injures a graduate student who loses fingers and eyesight.

June 13, 1985–UNABOMBER sends a letter bomb to the Boeing Aircraft Company. The IED is detected. The bomb is rendered safe.

November 15, 1985–UNABOMBER sends a letter bomb to the University of Michigan. A professor is injured when it explodes.

December 11, 1985–UNABOMBER places a disguised bomb in the parking lot of a computer rental store. The owner tries to remove the disguised device and is killed. This incident represents the first fatality from a UNABOMBER bomb.

February 27, 1987–UNABOMBER places a bomb disguised as a piece of lumber in a store parking lot. The store manager is severely injured when the bomb detonates as he attempts to remove the disguised device.

September 1988–An IRS office is bombed in California using bombs in a stolen vehicle parked in the building's garage.

February 1990–An attempted bombing of IRS office in California occurs. This occurs with a powerful VBIED, which fails to function. The device contains 2,000 pounds of alleged ANFO mixture in a stolen truck parked outside the IRS office.

February 26, 1993–A large bomb contained within a rental van is detonated under the World Trade Center in its underground parking deck. Six persons are killed and more than 1,000 injured. The van contained roughly 1,300 pounds of sensitized urea nitrate as the explosive. The device used a burning fuse delay and used aluminum, magnesium, and ferric oxide as sensitizers in the urea nitrate. The device is alleged to have used ammonium nitrate dynamite, nitroglycerin, and smokeless powder as a booster to detonate the main charge of urea nitrate.

April 1, 1993–A vehicle containing mortars fires 13 pipe bombs at an IRS office in California, nine explode, and several malfunction.

June 22, 1993–UNABOMBER sends a mail bomb to University of California. A geneticist is severely injured when the IED explodes.

June 24, 1994–UNABOMBER sends a mail bomb to Yale University. A computer science professor is seriously injured when the device explodes.

December 10, 1994–UNABOMBER sends a mail bomb to an advertising executive. The device explodes and kills him making him the second fatality in UNABOMBER's bombing campaign.

April 19, 1995–Timothy McVeigh uses a truck packed with nitromethane sensitized ammonium nitrate (AN) to destroy the Murrah federal building. The blast kills 169 people and injures more than 675. The bomb uses a burning delay and is estimated to be 4,000 pounds of sensitized AN. McVeigh uses a binary explosive along with commercial dynamite detonated by a blasting cap with a burning fuse dely. McVeigh uses AN in bags between the side of the rental truck and drums of sensitized AN to direct the blast into the building.

April 24, 1995–UNABOMBER kills his third victim with a mail bomb. The victim is a timber industry executive.

April 3, 1996–UNABOMBER is arrested at his Montana cabin. FBI finds extensive evidence linking him to the bombing campaign.

July 27, 1996–Centennial Park in Atlanta, Georgia, is the scene of a bombing. The device kills two and injures more than 100. The device is contained in a military ALICE field pack and consisted of three pipe bombs allegedly filled with smokeless gunpowder surrounded by nails to serve as enhanced fragments. The device weighs approximately 40 pounds. Steel plates are used to direct the blast and fragments toward the populated areas. Eric Rudolph is credited with the bombing.

January 16, 1997–A family planning clinic in Sandy Springs, Georgia, is bombed. The initial explosion damages an unoccupied examination room. The initial charge is placed on the back porch of the facility and appears to be nitroglycerin-based dynamite. The device is designed to lure public safety responders on scene where a second device buried in a flower bed will detonate. The second device detonates and injures at least seven individuals including law enforcement officers. A vehicle parked nearby absorbs some of the enhanced fragments which are nails. It appears a mechanical timer is used in the second device. The first device is believed to have used a burning fuse delay.

February 21, 1997–The Otherside Club in Atlanta, Georgia, is bombed allegedly by Rudolph. The bomb is secreted outside the club in a patio area where it exploded. Five individuals are injured. A second device is discovered by police and rendered safe. This device was contained in a similar knapsack as to the Centennial Park bombing. Mechanical timers in form of a windup clock are found in the devices.

January 20, 1998–Eric Rudolph is alleged to have bombed an abortion clinic in Birmingham, Alabama, killing a police officer and critically injuring a nurse. The device, which used nitroglycerin dynamite, was disguised as a potted plant and used nails as enhanced fragmentation. Some reports claim that Rudolph used a radio-controlled remote to detonate this device as Officer Sanderson of the Birmingham PD investigated the suspicious potted plant in front of the clinic. The ensuing blast killed Officer Sanderson and critically wounded Emily Lyons, a nurse who worked at the clinic. Unofficial reports claim that Rudolph also targeted a Birmingham Police facility with another bomb, but this device was rendered safe.

January 22, 1998–Theodore Kaczynski, UNABOMBER, pleads guilty to life in prison without parole in the bombings. He is placed in a super maximum security federal prison. He has killed three and injured 23 with 16 known bombs from 1978 through 1995.

August 7, 1998–Two VBIEDs are used to attack US embassies in Nairobi, Kenya, and Dar es Salaam, Tanzania, with the Nairobi bomb consisting reportedly of roughly 1,800 pounds of TNT packed in wooden boxes concealed in a box truck while the Dar es Salaam bomb used a similar amount of TNT with oxygen cylinders and ammonium nitrate with the blast directed by sand bags. The suicide bombers detonated the devices in proximity to the embassies, killing approximately 210 people in Nairobi while injuring about 4,600, with the Dar es Salaam bombing killing about 10 with 85 wounded. Most of those killed were local citizens.

April 20, 1999–Eric Harris and Dylan Klebold shoot and kill 12 students and one teacher, injured another 24 students prior to committing suicide. They brought with them several IEDs (99 IEDS were built) and emplaced several IEDs. Two 20-pound IEDs are placed in the school cafeteria. A firebomb was used as a distraction at another location but was extinguished by firefighters and failed to cause the distraction desired. They throw several pipe bombs and

Molotov cocktails of which some fail to function. Additional IEDs are found in the school parking lot in vehicles and the roof of the buildings. The filler for the devices was mainly flash powder from fireworks. The devices were of poor quality and only one injury resulted from their use.

2000 to 2014

July 7, 2005–The "7/7" bombings occur in London. Three IEDs explode contemporaneously and damage three underground trains. About an hour later, a double-decker bus is bombed. Some 52 persons are killed and more than 700 injured. These are PBIEDs contained with what is described as "rucksacks." The explosives are described as organic peroxides, likely TATP. Radicals supporting Islam are said to have been the bombers.

October 1, 2005–A student detonates a bomb on the campus of the University of Oklahoma near the stadium. The device was contained in a backpack and was likely composed of TATP. It appears that this event was a suicide as evidenced by a suicide note. No evidence could be found that the student had planned to enter the stadium.

July 11, 2006–In Mumbai, India, seven trains are bombed injuring more than 700 and killing more than 200. The attacks used cyclotrimethylenetrinitramine (RDX) mixed with sensitized ammonium nitrate. The explosives were contained in pressure cookers.

September 16, 2008–The US Embassy in Yemen is attacked with a VBIED which kills 16.

November 26, 2008–Multiple locations are attacked in Mumbai. Some 164 are killed and more than 300 persons injured. The 12 terrorists attack hotels, hospitals, and other sites over a several-day period. They use preplaced IEDs in vehicles to distract police. This is a classic swarm attack for which many agencies are ill prepared to respond.

December 25, 2009–A Nigerian citizen boards a flight from Paris to Detroit. As the flight approaches Detroit, the bomber spends 20 minutes in the restroom. He exits covered in a blanket and attempts to ignite the explosive by mixing two chemicals and adding them to the underwear bomb. Passengers see smoke and flame emanating from the underwear of the suspect and subdue the bomber. The bomber has a packet of PETN and triacetone triperoxide which he mixed with the two explosives and added a hypergolic mixture of ethylene glycol and potassium permanganate in an attempt to detonate the device. The device fails to explode. The bomber's genitals are burned.

May 1, 2010–A Pakistani builds a primitive bomb consisting of roughly 250 pounds of urea nitrate mixed with sugar, two gasoline cans with M88 fireworks attached, three propane cylinders, with travel alarm clocks as timers. The device ignites the M88 fireworks which fail to ignite the gasoline contained in the cans. The vehicle is spotted filled with smoke. The devices are rendered safe. Several reports state the urea nitrate–sugar mixture was not a functional explosive. The suspect is arrested leaving the US.

January 17, 2011–An alleged white supremacist places a pipe bomb configured as a mortar by welding one end closed which is loaded with more than 100 fishing weights coated with an anticoagulant. The device is placed along the Martin Luther King parade route. The device uses gunpowder with model rocket igniters and a remote radio controlled system to trigger the device. The device even though concealed in a backpack is discovered and rendered safe. The bomber is arrested.

April 15, 2013–Two IEDs explode during the Boston Marathon, at least three die and more than 170 are injured. Suspects are identified and one is killed during an attempt to arrest him. The second suspect is injured and later captured. The devices are alleged to be pressure cookers contained within backpacks. The explosive agent appears to have been flash powder from fireworks with enhanced fragmentation added in the form of nails and ball bearings.

Chapter 2

DEVELOPMENT OF THE BOMB SQUAD

Bomb squads have been in operation since the first professional unit was established in the Britain in 1874. Many from the US date to the early 1900s with New York City's bomb squad formed in 1903 and LAPD establishing a bomb squad in 1950. Miami-Dade County established a bomb squad in the early 1960s.

Photographs show commercial armor, albeit not a true bomb suit, worn as early as 1922. Even into the 1960s, bomb suits protected against fragmentation but did little to protect against overpressure. Many of the bomb suits covered only the front of the body.

However, many bomb squad members in these early years and even into the 1970s wore little or no protection and performed hand entries to render an IED safe. Such hand entries today are rare as robots and remote devices are used to manipulate suspect packages, or disrupters to open to package remotely. More commonly, the suspected IED is disrupted with bomb technicians a safe distance away. Professional bomb suits, which afford some degree of protection, are now the standard.

Older bomb squad operations emphasized either rendering the device safe by hand or moving the device. One sees many bomb technicians opening devices by hand and bomb transport containers in older photographs. These are both hand carried and vehicle transported. The first known bomb transport device was developed in Britain in 1890. Agencies hand built many of the bomb transport vessels. Many of these "homemade" transport devices were in use through the 1970s. Even today, some remain in service. Commercial

Figure 1. These photographs of the Milwaukee bomb truck were provided to Captain Brodie by former assistant police chief, Raymond Dahl. They are marked F. Schmidt Bureau of Identification, 3-12-41. Two principles of bomb protection are shown: distance and barriers. (From Thomas Brodie, *Bombs and Bombings*, 3rd ed. Courtesy of Charles C Thomas, Publisher, Ltd.)

Figure 2. Remote handling of a suspect bomb or improvised explosive device using distance as a protection factor. (From Thomas Brodie, *Bombs and Bombings*, 3rd ed. Courtesy of Charles C Thomas, Publisher, Ltd.)

Figure 3. The square corners of this bomb transporter were the weak points because the pressure of the explosion reflecting off the walls comes together at the corners. (From Thomas Brodie, *Bombs and Bombings*, 3rd ed. Courtesy of Charles C Thomas, Publisher, Ltd.)

transport devices are now available with some able to resist the explosion of 10 pounds of plastic explosives such as C4 (RDX explosive mixed with a plasticizer). The philosophy of moving a suspected bomb is no longer the primary method of operation but is still used on occasion (see Figures 1–3).

Early bomb squad personnel learned "on the job" and by the lead of more experienced bomb squad officers. The professional era of bomb squads began in 1971 when the FBI established Hazardous Devices School in Huntsville, Alabama. The FBI website says more than 20,000 bomb technicians have attended the program. Some states have additional standards for bomb technicians and bomb squads (see Figures 4–8).

Figure 4. Modern bomb suit.

Bomb Squads as a Resource

Bomb squads serve not only as a resource during a bomb or explosives-related event but also are useful in other settings. The ability to x-ray items has allowed agencies to examine firearms to determine if the firearm is loaded. This is helpful in cases of firearms found in a rusted condition.

In another instance, bomb technicians helped local businesses develop bomb threat and internal search protocols. Another agency surveyed local schools for explosive chemicals in their laboratory. They

Figure 5. This top view of the Brodie, Dykes and Zmuda bomb container shows the round cylinder. Following the instructions of Detective Bill Schmidt of the New York Bomb Squad, Brodie obtained an oil immersion tank and ten-weight motor oil for penetration and stopping mechanisms but never used it. It may be seen in the foreground. (From Thomas Brodie, *Bombs and Bombings*, 3rd ed. Courtesy of Charles C Thomas, Publisher, Ltd.)

found old containers of picric acid with the lids rusted onto the container, several pounds of metallic mercury, destroyed old containers of ethyl ether, and other chemicals labeled "explosive."

In concert with a hazardous materials disposal vendor and state department of environmental management, the bomb technicians developed a plan to move and dispose of the materials when the schools were not occupied. Several of the items were moved in bomb transport vessels, countercharged, and destroyed. The hazardous materials vendor took possession of the remainder of the material and disposed of it.

The bomb squad also supported the local intelligence efforts analyzing information regarding threat groups. The purchases of the

Figure 6. The latest model of the NABCO sphere is truly remarkable. The gases can be trapped inside for analysis and later released. Cost is approximately $240,000 and is being provided to all of the major airports in the United States by the Justice Department. (From Thomas Brodie, *Bombs and Bombings*, 3rd ed. Courtesy of Charles C Thomas, Publisher, Ltd.)

groups were assessed to determine if the group was attempting to manufacture explosives.

The bomb squad in some agencies assists in the safe storage and deployment of explosives such as breaching materials and distraction devices. Some agencies have integrated bomb technicians into their tactical teams and clandestine drug laboratory teams.

Bomb Squad Operations

The primary function of the bomb squad is to assess and render safe hazardous devices, typically bombs. In earlier times, the process was by attacking the object by hand entry or remote opening. The use of hand entry was the standard, but agencies rapidly saw the value of remote opening or movement of devices. The key is time, distance, and shielding. By using remote tools behind a shield, the device could be remotely opened and if detonation occurred, the officer was protected by shielding.

Figure 7. Thomas Brodie is shown finishing the dismantling of one of the most ingenious improvised explosive devices every encountered. This radio-controlled torpedo was designed to be used on a British ship carrying buses to Cuba. The can in the foreground was a shape charge container. The mechanism was tested in a lake before U.S. Customs seized it. Brodie said he learned a lot examining these seizures of over 4,000 bombs. (From Thomas Brodie, *Bombs and Bombings*, 3rd ed. Courtesy of Charles C Thomas, Publisher, Ltd.)

However, this is not feasible in some settings. The ability to x-ray items saw the use of medical fluoroscopes initially followed by portable x-ray equipment. The development of x-ray film onsite was the first evolution followed by line-linked computers to electronic receivers, to today's use of wireless receivers. In many instances, patrol officers and transit police investigate unattended items with x-ray equipment. If suspicious items such as wiring and batteries appear in the x-ray, the area is isolated, evacuated, and the bomb squad called. The ability to x-ray an item allowed the bomb technician to determine if the item was a bomb or appeared to be a bomb. This allowed precision disruption of the package (see Figures 9–12).

Figure 8. Miami-Dade County Police Technician Dominic Gerard is shown completing the dismantling of a combination improvised explosive and incendiary device. Several cans of gasoline all had explosives attached to them inside plastic bags and had detonating cord connecting them. The laundry business here had failed. (From Thomas Brodie, *Bombs and Bombings*, 3rd ed. Courtesy of Charles C Thomas, Publisher, Ltd.)

Early disrupters were shotguns using regular shot. As this technology developed, enterprising bomb technicians learned that water, wooden blocks, and frangible rounds could be used from the shotgun. The incompressible nature of water made it an excellent agent to attack the power source or switches of a suspected bomb in order to disable the device. Metallic pipes with end caps could be decapped using shotgun slugs, frangible clay rounds, or even steel rounds to penetrate heavy metal. This allowed the pipe bomb to be rendered safe without detonating in many cases. The shotgun was replaced by purpose made disrupters, which come with a variety of ammunition choices (see Figures 13–15).

As industrial robots were developed, they were adopted in the bomb setting. This provided the bomb technician the ability to use the robot cameras to assess the suspicious item, to move the item with manipulators if needed, or use a robot mounted shotgun or disrupter to attack the suspected device (see Figures 16–20).

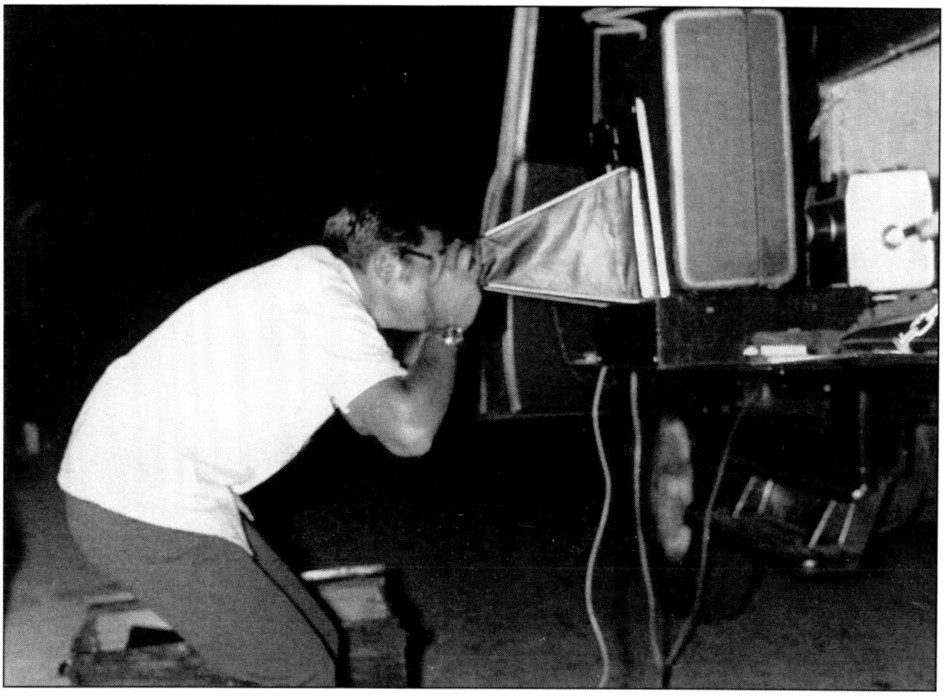

Figure 9. Captain Thomas Brodie, following Doctor Muhlberger's techniques, suspected bombs can be fluoroscoped with the x-ray machine and a lead-lined screen. This was an example of early x-ray equipment. The problem with this equipment was it placed the bomb technician close to suspect item. (From Thomas Brodie, *Bombs and Bombings*, 3rd ed. Courtesy of Charles C Thomas, Publisher, Ltd.)

In many instances, the robot is first downrange and officers in bomb suits only approach when the robot cannot complete the render safe procedure. Robots are now equipped with chemical sensors, radiation detection equipment, thermal imaging, a variety of manipulator claws, and multiple disrupters.

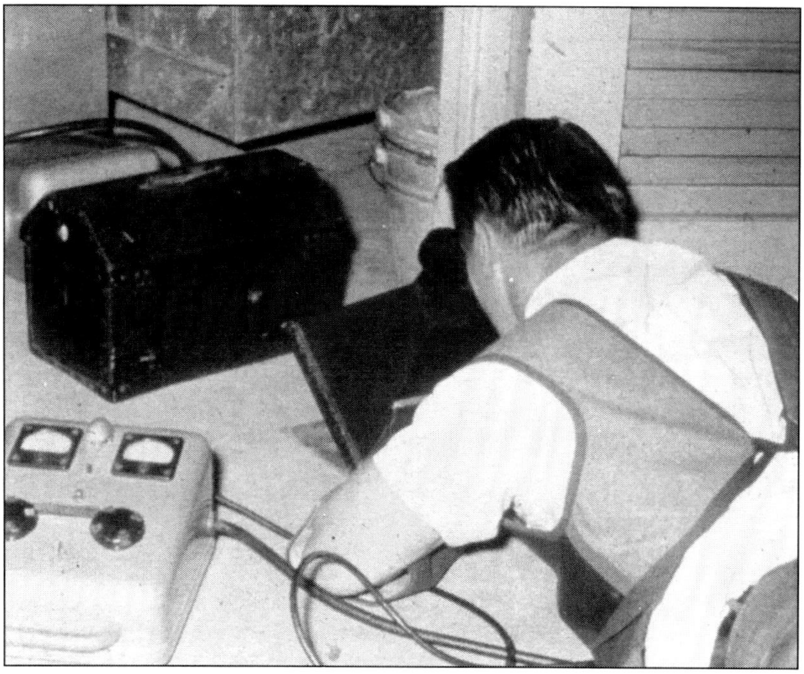

Figure 10. Captain Thomas Brodie is using lead and rubber gloves and apron to fluoroscope a suspected satchel in the Miami Police Department property room. This technique only requires one approach to a suspect bomb that has been handled. This technique places the bomb technician very close to the suspect device. (From Thomas Brodie, *Bombs and Bombings*, 3rd ed. Courtesy of Charles C Thomas, Publisher, Ltd.)

Figure 11. Here two suspect items are x-rayed using remote telemetry. The x-ray receiver plate has not been placed in this photograph. The items were unexpected and met sufficient criteria to warrant an x-ray. The items were detected by mailroom personnel who notified the author, as the packages had incorrect titles, wrong addresses, and were bulky, but more important, the items were not expected.

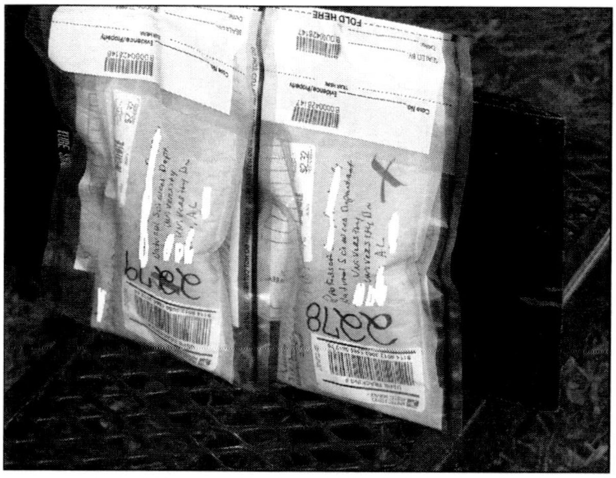

Figure 12. Here the x-ray receiver plate is in place behind the suspicious package. The items were promotional items not expected by the recipients. Bomb technicians were more than 100 feet away when they made multiple x-rays.

Figure 13. Bomb technician Corporal Chris Vida places a disrupter on a suspect package located near a gas meter. This was prior to the availability of a robot.

Figure 14. Officer Jason Youngblood (in bomb suit) and Officer Kevin McKee prepare to disrupt a suspicious suitcase. This was prior to robot availability.

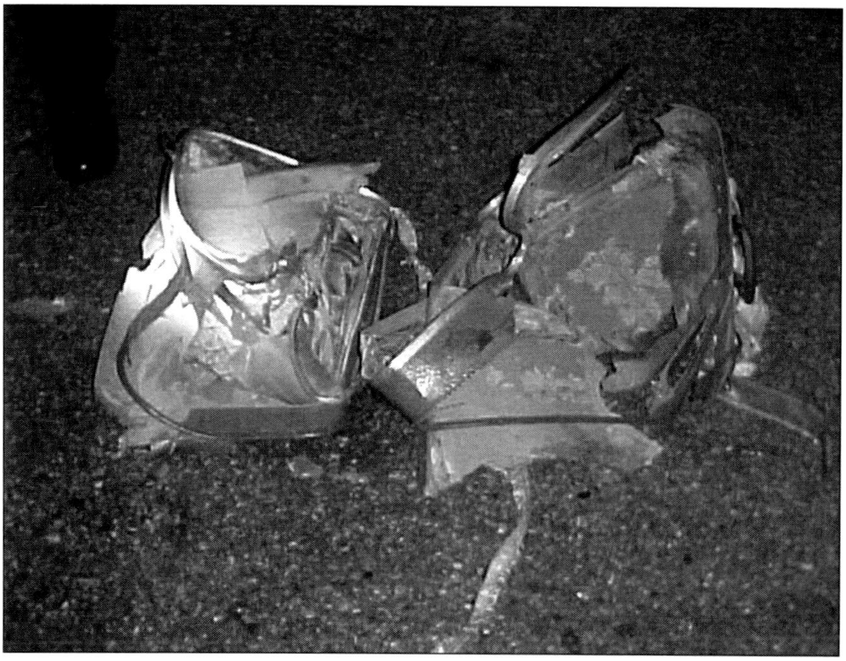

Figure 15. The suitcase following disruption by water shot. The suitcase was empty.

Figure 16. The use of firearms to dismantle bombs was one of the few options years ago that had erratic consequences when trying to deactivate bombs. (From Thomas Brodie, *Bombs and Bombings*, 3rd ed. Courtesy of Charles C Thomas, Publisher, Ltd.)

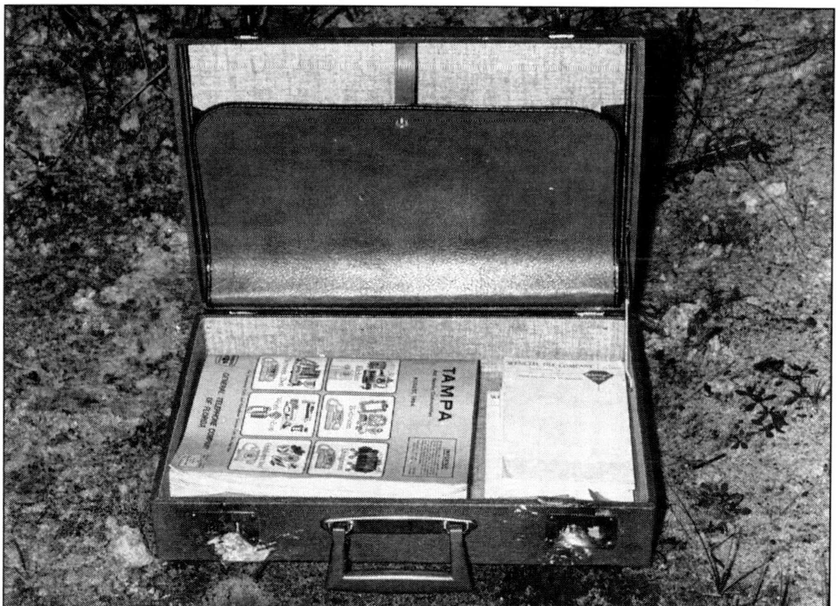

Figure 17. Before robots and disrupters, Thomas Brodie taped electric blasting caps to the latches of this briefcase, which was found at the Miami International Airport. The case was suspect because it had smoked. We found a book of burned paper matches inside the case. This technique requires touching of a suspect bomb but does not do much damage to the contents. (From Thomas Brodie, *Bombs and Bombings*, 3rd ed. Courtesy of Charles C Thomas, Publisher, Ltd.)

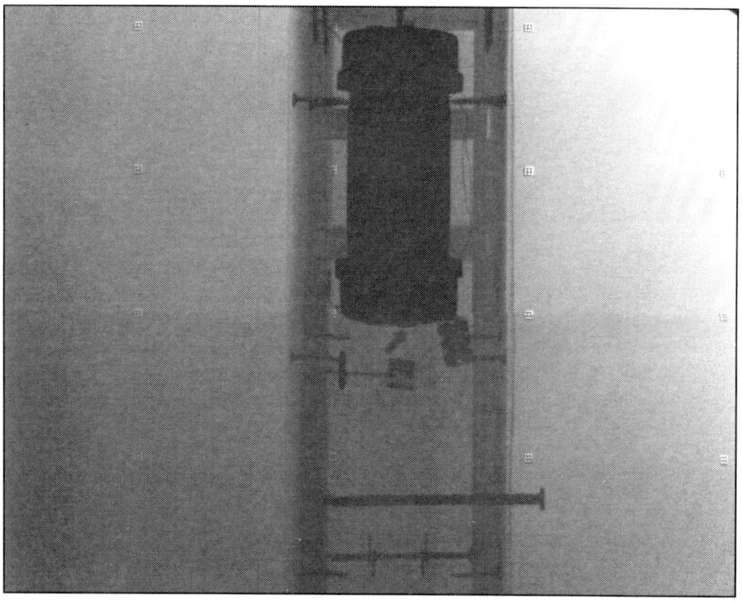

Figure 18. X-ray of a pipe bomb using a modern digital computer system.

Figure 19. Bomb technician and breacher, Corporal Chris Vida places a breaching charge on the door of a residence to be explosively breached.

Figure 20. Door following explosive breaching.

Chapter 3

EXPLOSIVES

Explosives are defined as any substance, which has the capacity to react or decompose rapidly with the release of energy. In many instances, the release of energy occurs over a few milliseconds. Heat, flame, gases, shock wave, and combustion products are usually released. In most instances, explosives are a combination of a fuel and an oxidizer. For the explosive to react or decompose, usually some form of energy is imparted to the substance, which prompts the combination of the fuel and oxidizer. This follows the traditional fire triangle although such occurs very rapidly.

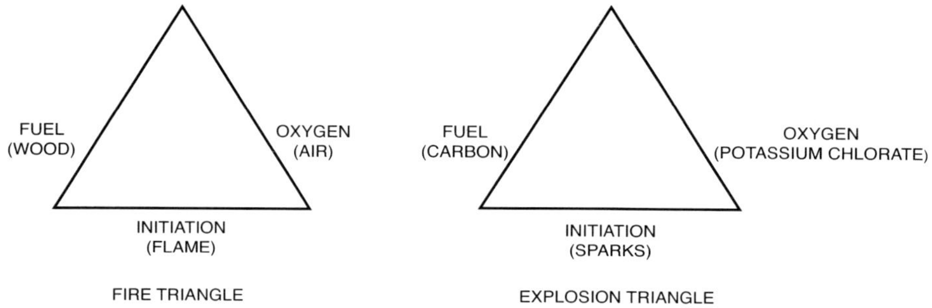

Figure 21. Simple fire triangle, well known to all firefighters, is applicable to explosives as well. The heat must be of sufficient temperature, sufficient duration, and cover sufficient area. Depending on the particular explosion, if these conditions are not correct, there may be a lot of narrow escapes. (From Thomas Brodie, *Bombs and Bombings*, 3rd ed. Courtesy of Charles C Thomas, Publisher, Ltd.)

Explosives typically contain a combination of carbon (C), hydrogen (H^2), nitrogen (N^2), and oxygen (O^2). These are called CHNO explosives. The byproducts in perfect reactions are carbon dioxide (CO^2), water (H_2O), and nitrogen (N^2). If incomplete combustion occurs, the following products may be produced: carbon monoxide (CO) and carbon soot. If oxygen is present in abundance then nitrogen oxides (NO^x) may be formed such as nitrous oxide (N^20). The amount of oxygen present intrinsically in explosives provides the underoxidized and are fuel rich or overoxidized reaction which is fuel lean (see Figure 21).

A hierarchy of reaction products is formed: first nitrogen forms nitrogen gas; second, oxygen and hydrogen form water; third, oxygen with carbon form carbon monoxide; fourth, any oxygen left forms carbon dioxide from carbon monoxide; fifth, any remaining oxygen forms oxygen gas; and sixth, traces, less than 1%, of nitrogen oxides are formed. If CHNO explosives contain any metal, they do not oxidize until the above reactions have occurred. Nitroglycerine is overoxidized and forms nitrogen gas, water, carbon dioxide, and oxygen.

Overoxidized explosives such as ammonium nitrate (AN), nitroglycerin (NG), produce gray smoke. They produce a smaller thermal event than do underoxidized explosives. However, TNT and RDX are underoxidized, thus fuel rich with lower levels of intrinsic oxygen. They produce dark smoke from carbon soot, carbon monoxide, and a secondary fireball is typically fueled from this mixture. This release of heat and subsequent fireball is called the thermal event. This is particularly true of underoxidized explosives such as TNT, which releases twice the heat in the secondary thermal event (fireball) as it does in the initial explosive event. Overoxidized explosives, tend to produce more gases than do underoxidized explosives. The gases tend to be carbon monoxide and carbon dioxide along with carbon soot, which provides the dark smoke from these explosions. Underoxidized explosives tend to be more toxic. These explosives are common in military applications and have high velocities of detonation. The higher velocity of detonation, the more shattering the effect (known as brisance) is of the explosive (see Table 1). The relative effectiveness is measured against the explosive power of TNT. A factor of one is equivalent to TNT (see Table 2).

Table 1. Velocity of Detonation (Approximate Values)

AN	7,600 fps to 15,000 fps (Used to attack Murrah federal building)
Black Powder	1,800 fps
DEGDN	20,000 fps
C4	24,000 fps
DNB	18,000 fps
Dynamite	15,000 to 21,000 fps
EGDN	21,000 fps
ETN	24,000 fps
HMTD	14,000 fps
HMX	28,000 fps
Lead azide	15,000 fps
Lead Styphnate	15,000 fps
PETN	24,000 fps
MEKP	16,000 fps
Mercury fulminate	13,000 fps
Nitroglycerin	23,000 fps
Urea Nitrate	20,000 fps (used to attack World Trade Center)
Picric acid	21,000 fps
RDX	26,000 fps
TATP	14,000 fps
TNB	21,000 fps
TNT	21,000 fps

Table 2. Relative Effectiveness Compared with TNT (Approximate Values)

AN	0.42 – 0.72
Black Powder	0.55
C4	1.34
DEGDN	1.10
DNB	0.60
Dynamite	0.80 – 1.25
EGDN	1.10
ETN	1.60
HMTD	0.80
HMX	1.7
PETN	1.60
Nitroglycerin	1.50
Nitrocellulose	1.10
Urea Nitrate	1.00
Picric acid	1.20
RDX	1.60
TATP	0.80
TNB	1.20
TNT	1.00

Figure 22. A binary charge with ammonium nitrate base explodes. Note the gray white smoke of an overoxidized explosive.

Figure 23. A car with 15 pounds of TNT placed in the front seat area. (Photograph by Sergeant Gary Coleman.)

Figure 24. The car explodes. Note the dark smoke on the outside of the fireball. This is characteristic of an underoxidized explosive such as TNT. (Photograph by Sergeant Gary Coleman.)

Figure 25. The remains of the vehicle following a 15-pound TNT explosion. (Photograph by Sergeant Gary Coleman.)

Overoxidized explosives, such as ANFO, tend to produce a large volume of gases, which make them useful for mining where gases do a substantial portion of the work moving material. However, the nitrogen oxides are corrosive, but overoxidized explosives still tend to be less toxic than the underoxidized explosives such as TNT or RDX (see Figures 23-25).

Pure explosives contribute energy when detonated and are neither a fuel nor an oxidizer. Lead azide is an example. The energy is released when high-energy molecular bonds are broken. Pure explosives release no byproducts and contain neither a fuel nor an oxidizer. The energy released comes from molecular bonds as they break. These tend to release large quantities of gases. In the case of lead azide, the nitrogen bond is broken allowing gaseous nitrogen to form. Another example of a pure explosive is TATP. Its decomposition products are acetone and oxygen.

Primary and Secondary Explosives, and Blasting Agents and Mixtures

Primary explosives are very sensitive to shock, flame, friction, heat, or movement. They will explode with little energy input. These explosives tend to be unstable.

Examples of primary explosives include:

- NG
- Lead azide
- Lead styphnate
- Mercury fulminate
- TATP
- Hexamethylene triperoxide diamine (HMTD)
- Diazodinitrophenol (DDNP)

Primary explosives are used in blasting caps and primers because of their sensitivity to flame or shock. This makes them suitable for triggering secondary explosives. Secondary explosives are not as sensitive as primary explosives and require more energy to start the explosion decomposition process. Usually blasting caps are required to initiate secondary explosives. Many secondary explosives will burn without exploding. However, fires in explosives should be treated as they will explode. Fires in explosives should not be fought but immediate evacuation is required.

Examples of secondary explosives include:

- PETN–used in boosters or detonation cord, blasting caps
- RDX–used in blasting caps, used in Composition C, Composition A
- TNT–used in combination with RDX in Composition B explosives
- Tetryl–2, 4, 6-Trinitrophenylmethylnitramine
- AN–ammonium nitrate when sensitized with nitromethane
- Binary explosives–usually AN sensitized with nitromethane

A common aromatic nitrogen-based explosive is trinitrobenzene (TNB), or dinitrobenzene (DNB). They serve as a base to manufacture other explosives such as TNT. Blasting agents are very insensitive to shock or flame. A blasting cap will not likely detonate these agents. Usually a booster of PETN or RDX is required to detonate blasting agents. Examples include AN, ANFO, and emulsion-based products used in mining. These tend to be overoxidized explosives.

Mixtures are agents, which are compounds, mixed, but are not chemically bound such as Tannerite®. Some of these agents are not explosives but are an incendiary pyrotechnic; agents sometimes classified as such include thermite and thermate. These agents use the Goldschmidt reaction to produce very high temperatures from the exothermic (heat-releasing) reaction. They use the chemical reaction of iron oxide and aluminum in an eight to three ratio weight based of iron oxide to aluminum power, along with an oxidizer such as barium perchlorate to produce molten iron slag at 2,000° F or higher as seen with commercial and military grade thermite mixtures. A magnesium strip ignited with a butane lighter or match is used in the improvised devices using thermite. Thermate adds barium nitrate and sulfur to raise the temperature to near 4,500° F. These agents are used in destruction systems and in some cases for welding (see Figure 26).

$$Fe_2O_3 + 2Al \rightarrow 2Fe + Al_2O_3 + \text{Heat}$$

Chemical Formula for a thermite reaction

Figure 26. Thermite burns through a one-half inch steel plate. Note the bright white flames and pyrotechnic particle emanating from the burn. (Photograph by Sergeant Gary Coleman.)

Black powder is a mixture as are some binary explosives such as Tannerite. This is a mixture of high-grade ammonium nitrate and aluminum powder. When struck with a bullet or initiated with a blasting cap, this mixture creates an impressive explosion. Several videos of these explosions are available on the Internet. Some of these items are sold over the counter. These compounds have been allegedly associated with several injuries or fatalities with ensuing litigation.

Fireworks contain flash powder, which is a mixture of a powdered metal and an oxidizer. Aluminum and magnesium are commonly used as powdered metal with potassium or sodium chlorate ($KClO_3$, $NaClO_3$). Some may add potassium nitrate (KNO_3) or potassium perchlorate ($KClO_4$) as oxidizers. Flash powders are extremely sensitive to friction, flame, and static electricity. Many unintentional ignitions in the manufacture and handling of flash powder occur because of static electricity. This makes homemade fireworks laboratories very dangerous for unintentional explosions.

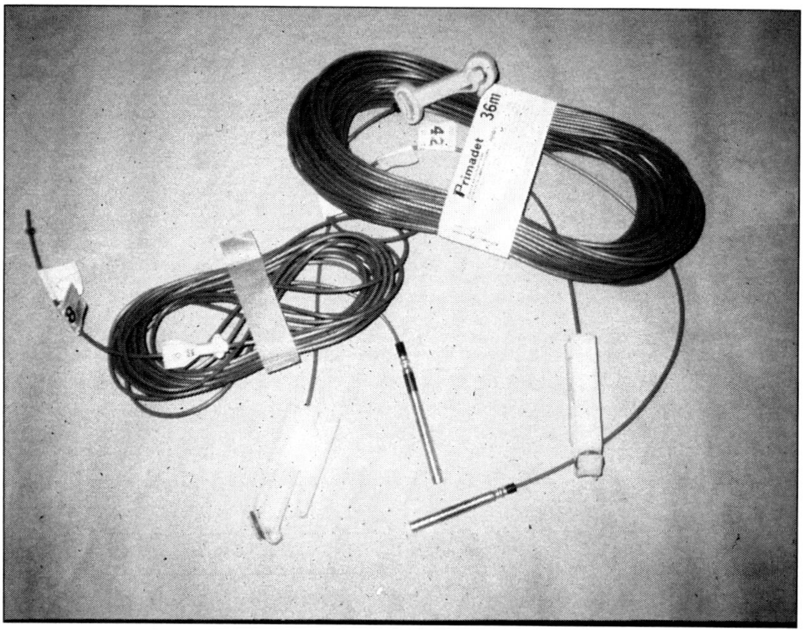

Figure 27. This is a nonelectric blasting cap. Note the shock tube connected to the blasting cap in lieu of metal leg wires seen in electric blasting caps.

Initiation of Explosives and Blasting Caps

Blasting caps were invented during the mid-1700s when Leyden jars (accumulators of static electricity) were used to ignite black powder. Primary explosives can be initiated by flame, impact, friction, or hypergolic reactions. Fuse-type blasting caps use a spit of flame to ignite a pyrotechnic mixture to ignite primary explosive contained within the cap. Electric matches have been used to ignite the primary explosive, exploding bridge wires, and electric filaments to ignite the primary explosive. The nonelectric (NONEL) blasting caps use a thin plastic tube, called shock tube, containing small amounts of an HMX (an RDX like explosive) aluminum mixture to propagate a percussive wave to the NONEL blasting cap. These are used widely in the blasting industry and by bomb squads as they eliminate the chance of stray electric currents inducing an explosion in an electric blasting cap. Blasting caps range from one to 10 on the Sellier and Bellot rating system. A number eight blasting cap will initiate many secondary explosives (see Figure 27).

Figure 28. These old electric blasting caps were found in a farm shed.

Blasting caps contain a base charge of a secondary explosive such as PETN or RDX in contact with the primary explosive to provide sufficient power to detonate a larger amount of a secondary explosive. Blasting caps have used mercury fulminate, lead styphnate, lead azide, and Tetryl as the primary explosives. More commonly seen is diazodinitrophenol (DDNP) and proprietary primary explosives. Blasting caps are sensitive to impact, friction, and flame. Blasting caps may degrade with age, producing peroxides, which make the blasting cap very sensitive to movement. Old blasting caps should not be moved or handled (see Figure 28).

Hypergolic reactions are flame, heat, and gases evolved from the combination of two substances at room temperature. Examples of hypergolic reactions are ethylene glycol and potassium permanganate, which the "Underwear" bomber tried to use to ignite TATP contained in his underwear in his failed attempt to bring the airliner down. Other organics have hypergolic reactions such as brake fluid and high-test hypochlorite (HTH) used in pools, or sulfuric acid mixed with potassium chlorate and table sugar.

Energy Released, Work, and Damage

Explosives use two mechanisms of energy release when high explosives react. Extremely rapid oxidation or rapid decomposition produces a shock wave or shock energy wave, heated gases, or gas energy. Low explosives typically produce little or no shock energy with their effects produced by gas energy. Low explosive velocities typically are less than 3,300 feet per second. Examples of low explosives include smokeless gunpowder, black powder, and nitrocellulose. These substances deflagrate or burn at a rapid rate but do not explode unless confined. They do produce hot gases, which can cause container failure and an explosion. Low explosives do not produce a transonic shock wave. The typical pressures generate are low overpressure in the 150,000 psig range.

The density of the explosive material affects the amount of energy imparted. As a rule, the denser the explosive, the more energy produced. However, if one compresses an explosive too much, called "dead pressing," the explosive tends to be desensitized and may not explode. Small bubbles of air are required for most explosives to function. Some blasting agents are aerated while others may have microspheres of glass inserted during manufacture to improve sensitivity. When an explosive is initiated, the area of rapid oxidation and shock wave proceed along the body of the explosive. If the explosive is in a cylindrical form, little radial energy is released. To propagate the explosive wave, a certain diameter of explosive is required. This is called critical diameter. Dynamites typically have a critical diameter of one inch while AN may vary from seven to 15 inches depending upon the sensitizing agent used.

As the explosive is ignited, the initial ignition is deflagration, which rapidly changes to a supersonic detonation wave in high explosives. The wave is called the Chapman-Jouget plane, which moves along the body of the explosive. Explosive velocities can be increased to improve their brisance or shattering ability by increasing the density of the explosive without dead pressing, or decreasing particle size in AN, or in the case of low explosives, increasing the confinement. In some cases, low explosives can be initiated with a blasting cap to increase the velocity of reaction.

Other types of explosions include mechanical explosions in which energy causes the failure of a container. An example is the failure of a steam boiler vessel or steam generated in concrete, which causes spalling. Spalling can be differentiated from explosions caused by high explosives as the crater edges in the concrete will be irregular and have sharp fracture lines, whereas those from high explosives do not.

High explosives do not have to be confined to explode and they have detonation velocities of more than 3,300 feet per second. High explosives produce a transonic shock wave, which precedes the hot gases produced, and thermal event. The shock wave produces an immediate pressure spike, which may be more than 1,000,000 psig. As the shock wave displaces the atmosphere during its passage, a secondary and much weaker but longer reverse or negative shock wave returns to fill the lower atmospheric bubble created by the explosion. Common detonation velocities are 15,000 feet per second with some even faster. These higher velocities of detonation make the explosion more shattering or brisant. This imparts more energy and tends to shatter the target rather than "push it with gases" as is seen with low explosives and overoxidized high explosives.

Shock waves are short-range phenomena but may be reflected by solid surfaces or even nonsubstantial surfaces. Bags of water, mud, newspaper, and even clouds can reflect shock waves. The reflected shock waves may produce unusual injury or damage patterns. These reflections are the reason water is used to buffer explosions used in some breaching applications. The water directs the shock wave in the direction needed without creating dangerous fragments. This technique can use other materials such as steel plates or environmental features to control the direction of the shock wave and along with it a substantial amount of the fragments produced by the explosions.

The damage done to a target by the shock wave is about 10–15% of the total energy liberated. The gas energy tends to account for about 85–95% of the energy. Shock waves dissipate rapidly and have a short range whereas gases travel further. However, objects from the explosive container, or added enhanced fragmentation such as nails or screws, and objects "picked up" by the explosion may travel much further. This effect regarding gases makes overoxidized explosives such as ANFO the choice for mining as its shock wave shatters the material and the gases push it off the bench making it accessible for equipment to transport. ANFO is also an optimal explosive to attack buildings, create craters, and for moving rock as the gases from the explosion tend to "push" the structure causing structural failure.

The work accomplished by an explosion is affected by the area impinged upon by the explosion. This means the more explosive in contact with the target, the more effective the energy transfer will be. Even a few inches of air gap will dramatically decrease the damage done by an explosion. For this reason, increasing the distance, standoff distance, can be critical for protection. Shock waves in many cases will have multiple reflections in most environments providing multiple peaks in pressure. Shock waves tend to reflect at a 180° angle. In many instances, a shock wave will flow around objects creating air vortices, especially 90° to the object encountered. The sharper the surface encountered, the more substantial the vortices will be. Shock waves tend to follow the inverse square law, which means the shock wave loses energy rapidly and is a short-range phenomenon. These vortices may account for shock wave damage on the sides of objects, which do not face the blast source. Damage on the rear side of such objects is not uncommon in large explosions. Another postulated mechanism of damage is at interface points where the shock wave transitions from one media to another. An example would be spalling on the inside of a structure's wall when the blast wave transitions from the exterior of the wall to the interior of the room. This change in speed of propagation is postulated to release energy and a spalling of material could be a result from such.

As shock waves and gases penetrate a structure, they may cause "drag" on the interior structure walls, floor, and overhead causing structural collapse. Shock waves are transmitted via the ground and may cause damage from ground movement. This may also contribute to structural collapse. Usually very large devices such as VBIEDs must be used to cause structural collapse from these mechanisms.

Special Hazards

History abounds with stories of unintentional explosions. Some of the more substantial unintentional explosions are examined along with special hazards presented by some explosives or explosive materials, booby traps, toxicity issues, and placarding of explosives.

Figure 29. This 34-0-0 bomb grade fertilizer was seized from a bomb maker. His target was the police department. The suspect complained to officers when they would not return his "bomb-making material." The suspect even provided a written statement to the police chief regarding the officers' actions.

Unintentional Explosions

Texas City Ammonium Nitrate Fertilizer Explosion

One of more substantial events involving AN is the April 16, 1947 fire and explosion in Texas City, Texas. The explosion resulted from a fire aboard a ship transporting ammonium nitrate fertilizer. Local firefighters responded with equipment and were fighting the fire. The shipboard effort of introducing live steam into the hold was fruitless as ammonium nitrate is an oxidizer and supported the combustion. The firefighters and ship's crew ignored the bulging of the sides of the ship and boiling off of water touching the ships hold, along with the ominous sign of reddish brown smoke. Approximately, 2,280 tons of ammonium nitrate exploded (see Figure 29).

The resulting explosion broke windows 25 miles away and was heard 150 miles away. The explosion produced a 15-foot wave, which sank nearby ships, started a fire on another ship also carrying 961 tons of ammonium nitrate fertilizer, and tore the wings off two sightseeing aircraft flying overhead causing them to crash. Almost the entire Texas

City Fire Department, 27 out of 28 members, was killed and its four pieces of fire apparatus were destroyed. Nearby structures disintegrated and objects weighing almost 3,000 pounds were thrown 13,000 feet from the explosion site. Efforts to salvage the ship set fire near the original explosion, which contained 961 tons of AN fertilizer, unsuccessful and it also detonated.

The destruction included four ships sunk, total destruction within a one mile radius, 125 oil storage tanks damaged or destroyed, 600 vehicles destroyed, and 360 rail cars damaged or destroyed. More than 1,000 structures were damaged or destroyed. The casualties included 468 killed and more than 2,000 injured. Some 113 of those present had no bodies recovered.

West Texas Fertilizer Plant Explosion

In 2013, a fertilizer plant manufacturing ammonium nitrate fertilizer caught fire and approximately 30 tons of ammonium nitrate fertilizer detonated. Emergency services personnel were on scene fighting the fire when the detonation occurred. Ten first responders were killed in the explosion along with several persons outside the facility for a total of 15 fatalities. Some 200 persons were injured and 150 buildings were damaged or destroyed. About 20 minutes elapsed from the time the fire department was called until detonation. It appears that first responders were unaware that ammonium nitrate was involved in the fire. Other ammonium nitrate explosions are shown in the Table 3.

Transportation Incidents

Incidents involving an explosion in the transportation of explosives are not common. However, a recurring theme is the lack of protective measures such as evacuation when fire involves the cargo of explosives. A classic case of fire causing detonation of explosives is the 1959 Roseburg, Oregon event. A truck transporting two tons of dynamite and roughly four tons ammonium nitrate blasting agent became involved after the building adjacent to the vehicle caught fire. The ensuing explosion created a crater 52 feet wide and roughly 20 feet deep. Responding police officers and firefighters did not note the explosives placards on the truck. The explosion killed 14 and destroyed every

Table 3. Other Ammonium Nitrate Explosions

Date	*Location*	*Deaths*	*Details*
1823	Germany	1,300	AN plant explodes
1916	United Kingdom	120	150 tons AN exploded at an ammunition factory
1918	US	>100	21 million pounds of AN detonate
1921	Germany	>500	Explosives used to break up AN piles, 30 tons exploded
1942	Belgium	189	Explosives used to break up AN piles, 150 tons exploded
1947	France	29	Shipboard fire with 3,300 tons of AN exploded
1988	US	6	50,000 pounds of AN blasting agent exploded killing the firefighters present
1994	US	4	Fertilizer plant exploded, 5,700 tons of AN involved
1998	China	22	Fertilizer plant exploded, 27 tons of AN involved
2001	France	31	Warehouse with 200 tons of AN exploded, more than 2,000 injured
2004	North Korea	162	Train carrying AN explodes, injures more than 3,000
2007	Mexico	40	Truck wreck, 22 tons AN explode
2013	US	15	Fertilizer plant catches fire and explodes, more than 200 injured

building within an eight-block area. Structures within a 30-block area were damaged heavily. At least 125 others were injured.

Fragments from the blast killed many and broken glass caused the death of at least two individuals. At least three of the firefighters on scene were literally blown apart as was at least one police officer. This incident caused sufficient consternation as to cause a change in many explosive transportation laws. A study by the NIOSH provided information on several transportation accidents in the US. In 1998 in Walden, Ontario, a truck transporting 40,000 pounds of blasting agent caught fire following the truck crashing. The area was evacuated and firefighters did not fight the fire. The truck exploded roughly 30 minutes into the fire and projected fragments 5,800 feet. Two minor injuries were reported.

A truck carrying 35,500 pounds crashed in Utah and caught fire. The driver and co-driver were evacuated by other motorists and were trying to leave the scene when the truck exploded. The explosion created a 30-foot deep crater 70 feet wide. Some 17 persons were injured.

Ammonium Nitrate

Although high-grade ammonium nitrate fertilizer is not as common because of Homeland Security regulations, it still warrants discussion. As noted in Table 3, the involvement of explosions with ammonium nitrate fertilizer and blasting agents occurs on an alarming basis. The minimum evacuation distance for any sizeable amount of ammonium nitrate fertilizer on fire is 5,000 feet.

In many of the circumstances, a fire involving the ammonium nitrate fertilizer or blasting agent is not recognized as a potentially explosive event. Ammonium nitrate fertilizer is placarded as an oxidizer. This is misleading. Ammonium nitrate fertilizer in the 34-0-0 grade is literally bomb-grade material.

Ammonium nitrate is manufactured by combining ammonia and nitric acid. When ammonium nitrate is heated above 300° F, it begins physical and chemical changes, which make it unstable. In an exothermic reaction, it decomposes to nitrous oxide and water with a white-colored smoke. Ammonium nitrate can have its detonation potential increased by contamination with almost any organic material or metals. As it melts, it may flow from its container and interact with soil, fuels, or metals making it more susceptible to detonation. At around 575° F, it begins to decompose rapidly into nitric oxides, which produces a reddish-brown smoke. Remember that ammonium nitrate involved in a fire can explode at any point, but if reddish-brown smoke is visible, the prudent action is to presume an explosion is imminent. Fires in which the fertilizer is contained, such as the holds of ships or rail cars are particularly dangerous as they concentrate heat and accelerate the decomposition of the ammonium nitrate.

The more dangerous form of ammonium nitrate fertilizer is in the prilled form which is coated with wax. The wax provides an organic

ATF	VEHICLE DESCRIPTION	MAXIMUM EXPLOSIVES CAPACITY	LETHAL AIR BLAST RANGE	MINIMUM EVACUATION DISTANCE	FALLING GLASS HAZARD
	COMPACT SEDAN	500 Pounds 227 Kilos *(In Trunk)*	100 Feet 30 Meters	1,500 Feet 457 Meters	1,250 Feet 381 Meters
	FULL SIZE SEDAN	1,000 Pounds 455 Kilos *(In Trunk)*	125 Feet 38 Meters	1,750 Feet 534 Meters	1,750 Feet 534 Meters
	PASSENGER VAN OR CARGO VAN	4,000 Pounds 1,818 Kilos	200 Feet 61 Meters	2,750 Feet 838 Meters	2,750 Feet 838 Meters
	SMALL BOX VAN (14 FT BOX)	10,000 Pounds 4,545 Kilos	300 Feet 91 Meters	3,750 Feet 1,143 Meters	3,750 Feet 1,143 Meters
	BOX VAN OR WATER/FUEL TRUCK	30,000 Pounds 13,636 Kilos	450 Feet 137 Meters	6,500 Feet 1,982 Meters	6,500 Feet 1,982 Meters
	SEMI-TRAILER	60,000 Pounds 27,273 Kilos	600 Feet 183 Meters	7,000 Feet 2,134 Meters	7,000 Feet 2,134 Meters

Figure 30. This chart shows the recommended evacuation distances based upon the amount of explosives present.

substance to sensitize the ammonium nitrate. Some reports have stated that when exposed to temperatures in the 800° F range, it may detonate. Fires should be fought defensively with unmanned appliances with the object to cool the ammonium nitrate fertilizer. Remember that ammonium nitrate is a potent explosive. If ammonium nitrate-based blasting agents are involved in a fire, the fire should not be fought but evacuation undertaken using the North American Emergency Response Guide NAERG or WISER for minimum evacuation distances. Remember that ammonium nitrate fertilizer behaves in an unpredictable fashion (see Figure 30).

Anhydrous Ammonia

Although labeled as a nonflammable gas, anhydrous ammonia is still a substantial hazard and under the appropriate circumstances is explosive. Anhydrous ammonia is usually stored in pressurized containers, which range from a few pounds to multiton quantities in rail cars. Anhydrous ammonia is stored commonly in large bulk quantities in tanks or in towed tanks used in the agricultural setting. Fixed re-

Figure 31. The anhydrous ammonia placard, even though marked as non-flammable, will explode if confined with a "fuel rich" setting and high temperature ignition source.

frigeration systems may use anhydrous ammonia such, as refrigerated warehouses, ice cream producers, and other industry.

The theft of anhydrous ammonia for use in the manufacture of methamphetamine is an issue in many areas. In some cases, the theft may cause a failure of valves. One aspect one must consider is anhydrous ammonia is a very serious inhalation threat and to a lesser extent, a corrosive hazard. Contact with the cryogenic liquid, although rare, may cause serious injuries. Inhalation injuries are fatal in many events.

The United Sates Department of Transportation classifies anhydrous ammonia as a nonflammable gas in Class 2.2. The placard carries a green background with the drawing of a compressed gas cylinder on it and the four-digit code "1005" (see Figure 31).

Even though labeled as nonflammable, anhydrous ammonia, if confined and exposed to an ignition source, anhydrous ammonia will explode. Its lower explosive limit (LEL) is listed as 15% and its upper explosive limit (UEL) is listed as 28% in the National Institute of Occupational Safety and Health (NIOSH) handbook. This makes leaks of anhydrous ammonia in structures a very serious explosion hazard. Historical reports of such explosions exist and in many instances occurred in plants manufacturing ice cream or processing other foods requiring refrigeration systems. Leaks from the systems occur when piping is disrupted or pumps overheat.

Common Homemade Explosives

TATP and other peroxides are easy to manufacture. First discovered in 1896, this explosive has become popular as an illegal explosive because of its ease to manufacture. TATP requires acetone, at least 30% hydrogen peroxide (hair bleaching product or swimming pool product), and hydrochloric acid. These items are over-the-counter items easily acquired. TATP appears as a white to off-white powder, which may be found in a refrigerator to slow decomposition. It begins to decompose immediately after manufacture and has the odor of bleach. The more common form is tricyclic acetone peroxide (TCAP). TATP may form oligomers, which are molecules in the form of monomers, dimers, or trimers of TATP. The trimer form is the more stable of the oligomers. TATP is one of the few primary explosives, which does not contain nitrogen. Many explosive detection systems key on nitrogen as an indicator of explosives present. Small quantities of unconfined TATP may burn, but in quantities more than four to five grams, it will likely explode. Any confinement will cause an explosion when TATP is ignited. Peroxides operate in a different manner than most explosives. Explosives typically contain a fuel and an oxidizer, whereas TATP and other peroxides have high-energy bonds. When these bonds are broken, the TATP decomposes to gas at very high rate producing the explosion. The decomposition produces little heat and thermal event. The decomposition is an entropy burst at supersonic speed producing a tremendous amount of gas. TATP, sometimes called the "Mother of Satan," is very friction, shock, and heat sensitive.

Some individuals keep the reaction vessel in a refrigerator to emphasize the production of the trimer form, which is supposedly (at least marginally) more stable than the monomer and cyclic forms of TATP. TATP is a primary explosive as is its precursor diacetone diperoxide (DADP). It usually appears as a white or off-white powder, which might be confused with methamphetamine. One of the first reported illicit manufacture of TATP in the US was in 1983 when a HME exploded in California. In 1989, a postblast investigation in Maryland revealed a pipe bomb had killed four juveniles. The pipe bomb was an

Figure 32. Homemade TATP that was manufactured by a juvenile. The TATP is contained in the glass jar in the center of the photograph. This was one of several explosives the juvenile had manufactured. He took flash powder and TATP to school.

ammonium nitrate and aluminum flakes combination, but search warrants revealed TATP and HMTD had also been manufactured. The discovery of TATP and HMTD has been ongoing. Field drug test kits if used to test TATP may initiate an explosion. The author has seen attempts and successful manufacture of TATP. If a HME laboratory is discovered, one should presume that TATP is present. The velocity of detonation of TATP is roughly 14,000 fps and its relative effectiveness when compared with TNT is 0.8 (see Figure 32).

Hexamethylene Triperoxide Diamine (HMTD)

Hexamethylene triperoxide diamine (HMTD) is peroxide, which is easily manufactured using hexamine from solid fuel bars; at least 30% peroxide, and dilute citric acid or sulfuric acid. HMTD, like TATP, is very sensitive to friction, shock, and heat. HTMD is marginally more stable than TATP. HMTD was synthesized in 1885. HMTD is a primary explosive and may appear as a white or off-white powder.

HMTD may smell like dead fish or have a fishy odor as it decomposes but fresh product is odorless. This product may react violently when in contact with metals. Field drug test kits, if used to test TATP or HMTD, may initiate an explosion. Its velocity of detonation is approximately 14,000 fps and its relative effectiveness compared to TNT is 0.8.

Methyl Ethyl Ketone Peroxide (MEKP)

This explosive is manufactured by mixing methyl ethyl ketone with at least a 30% peroxide and slowly adding concentrated sulfuric acid. MEKP is marginally more stable than TATP. MEKP was synthesized in the early 1900s. The temperature must be controlled and the explosive is an oily liquid. This, like other peroxides, is sensitive to shock, heat, and flame. In some instances, expedient explosives can be manufactured by mixing peroxides with nitromethane or ethanol. These are very sensitive to heat, friction, and may even self-heat by decomposition and ignite. Organic materials may be added as a fuel, which might include flour or even coffee. MEKP has a detonation velocity of more than 15,000 fps.

Erythritol Tetranitrate

Erythritol tetranitrate (ETN) is an easily made explosive related to Pentaerythritol tetranitrate (PETN). ETN is more unstable and sensitive to shock, friction, and heat then PETN. ETN is manufactured by nitrating erythritol, a sugar, by mixing sulfuric acid and ammonium nitrate or potassium nitrate with erythritol. It may appear as a white or off-white powder with no odor. Its velocity of detonation approaches 24,000 fps. ETN has a long shelf life and may retain much of its explosive power some three to four years following manufacture. ETN is an overoxidized explosive with a positive oxygen balance as opposed to PETN which has a negative oxygen balance.

Nitroglycerin and Ethylene Glycol Dinitrate

Nitroglycerin and Ethylene Glycol Dinitrate (EGDN) are manufactured by nitrating glycerol in the presence of sulfuric acid. EGDN is manufactured by the same process except ethylene glycol is substituted for glycerol. Otherwise, EGDN behaves in a similar manner to NG. This is a very temperature sensitive process which requires a cooling bath. It produces toxic fumes and is a very dangerous exothermic process. This process is subject to thermal runaway with a resultant explosion. This clear to maroon oily liquid is very unstable. It will explode with shock, heat, flame, or friction. It may be desensitized by mixing with ethanol or acetone but remains explosive. Darker red colors indicate the formation of unstable compounds within the nitroglycerine.

TRINITROGLYCERIN (NITROGLYCERIN)–Thick oily liquid, shock sensitive, crystallizes at 56° F, crystals less dangerous than liquid, liquid may be milky or yellow, very shock sensitive when thawing, explodes around 500° F, hammer blow will detonate (8 lb. impact), emits poisonous gases when burning, liquid is poisonous, skin absorption, very brisant, high explosive.

Nitrogen tri-iodide and nitrogen trichloride are very sensitive explosives. They are made by combining ammonia and iodine. Any friction or shock will detonate this releasing a cloud of toxic iodine gas and particles. This explosive is so sensitive that in some cases, the simple process of crystallization may detonate the explosive. Nitrogen trichloride, a rare explosive, is synthesized by using anhydrous ammonia or ammonium nitrate by bubbling chlorine through either of these. Ammonium chloride may be used with battery electrodes immersed in the liquid with power flowing through generates this product. It appears as a yellow liquid. This explosive is very shock sensitive.

Mercury Fulminate

Mercury fulminate is a primary explosive, which has been used in blasting caps and percussion caps. Mercury fulminate is manufactured by dissolving metallic mercury in nitric acid and adding ethanol.

Black Powder, Flash Powder, and Chlorate Explosives

Black powder is a combination of sulfur, potassium nitrate, and charcoal. Some processes use potassium or sodium chlorate as an added compound. Flash powder is manufactured using powdered metals such as aluminum or magnesium mixed with potassium chlorate or sodium chlorate. These are very friction and flame sensitive. Static electricity may ignite them.

Armstrong's reagent is a combination of red phosphorus and potassium or sodium chlorate mixed with an organic solvent. This mixture is so unstable that even moving the red phosphorus container or stirring the red phosphorus and potassium chlorate mixture may cause an explosion. Red phosphorus is a very motion and friction sensitive material and will ignite easily. The explosion of Armstrong's reagent produces a large thermal event.

TNT and Related Compounds

TNT is manufactured by nitrating toluene with nitric and sulfuric acid to mononitrotoluene (MNT), with continued nitration to dinitrotoluene (DNT), and finally nitrated to TNT. This is a very temperature sensitive process in which, heat must be removed and the temperature controlled carefully. The same process can be used to make benzene-based explosives such as mononitrobenzine, dinitrobenzene, and trinitrobenzene. TNT is shock and heat insensitive, and appears straw yellow to pale yellow. TNT melts at 178° F and may be shock and heat sensitive if it appears brown. TNT exposed to heat above 350° F may cause detonation and makes the TNT shock sensitive. TNT becomes sensitive as it ages. Some commercial forms of TNT include Nitopel (pelletized TNT); TNT-based explosives: Amatex, Amatol, Ammonal, Baronal, Barratol, Boracitol, Compositon B, Compostion B-2, Composition B-3, Cyclotol, DBX, Minex, Ednatol, H-6, HBX, HTA, Minol, Octol, Pentolite, Picratol, Plumbatol, PTX-1, PTX-2, Sodatol, Tertrytol, Trinotal, Torpex.

Picric acid is also known as 2, 4, 6, trinitrophenol or ammonium picrate, and is one of the first nitrated explosives using phenol. As long as picric acid is wet, more than 10% water by weight, it remains stable. However, once dry or in contact with metal or organic material, it becomes very unstable. In earlier times, picric acid was used to clean glassware in chemical laboratories. It may still be found in some laboratories. Some World War II munitions use picric acid as filler. Japanese pottery grenades are sometimes found containing picric acid or an ammonium perchlorate mixture. These become very unstable with age. These grenades or containers with dried picric acid become very shock sensitive as are containers with metal lids. Several individuals have been injured handling Japanese pottery grenades or in attempts to remove the lids from dried containers of picric acid. These items are still found in some medical and academic laboratories. The method of disposal is movement in a blast resistant container and the use of a large charge of C4 or other suitable explosive is used to consume the picric acid. Issues with RCRA may be encountered in this disposal process with consultation with local environmental authorities a necessity

A common homemade explosive sometimes used as solid rocket fuel is potassium nitrate mixed with confectioners' sugar or table sugar. The sugar serves as the fuel while the potassium nitrate is the oxidizer. This combination will ignite when exposed to flame or sulfuric acid. It burns at a high temperature with white-hot flames. The mixture, if confined, will explode. Multiple sources on the Internet provide video and detailed instructions on its manufacture. Some individuals "cook" this mixture on residential stoves. The mixture is temperature sensitive and can ignite easily if overheated (see Figures 33–36).

Figure 33. This is a rocket fuel deflagration during cooking on stove. This combination of potassium nitrate and sugar ignited. It did not explode, as it was not confined. Note the burned rocket fuel on the stove.

Figure 34. Note the splattering and charring from the rocket fuel ignition.

Figure 35. These are the molds into which is poured the molten rocket fuel mixture. This allows the rocket fuel to be inserted into the rocket body.

Figure 36. Note the charred papers, which contained the directions for manufacturing the rocket fuel.

Homemade Explosive (HME) Laboratories

The problem with homemade explosive laboratories is they may resemble methamphetamine laboratories, other illicit drug laboratories, or biological labs manufacturing such items as ricin. In open source materials, the US Bomb Data Center (USBDC) cites about 420 cases per year involving homemade explosives or roughly, 24% of the bomb related incidents occurring in the US (see Figures 37 & 38).

Figure 37. A solid rocket motor is shown that is ready for insertion into the rocket body.

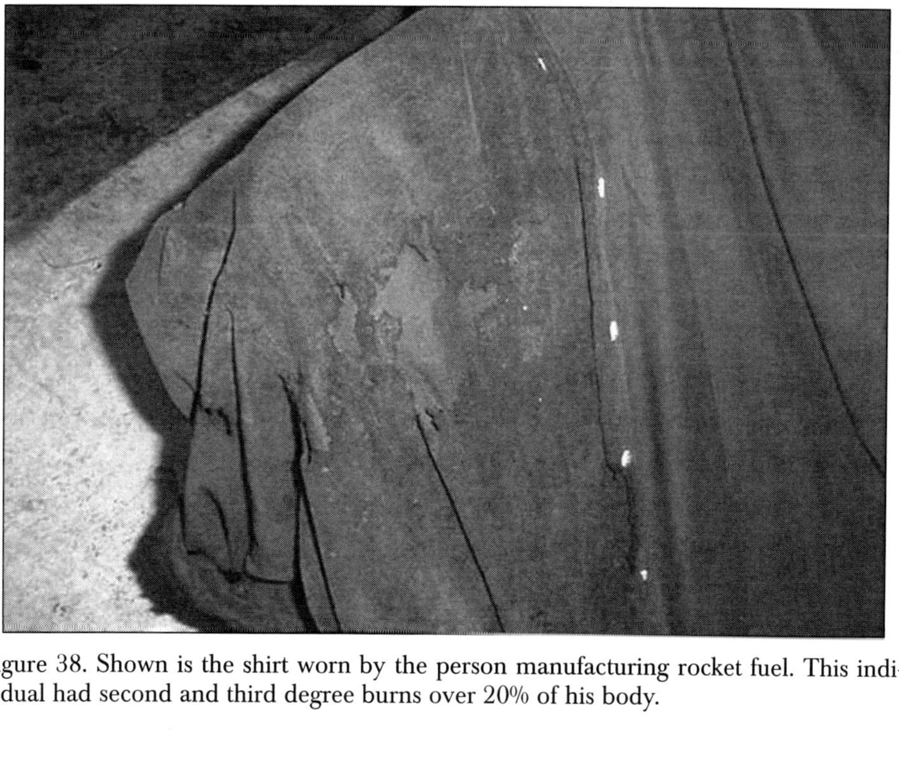

Figure 38. Shown is the shirt worn by the person manufacturing rocket fuel. This individual had second and third degree burns over 20% of his body.

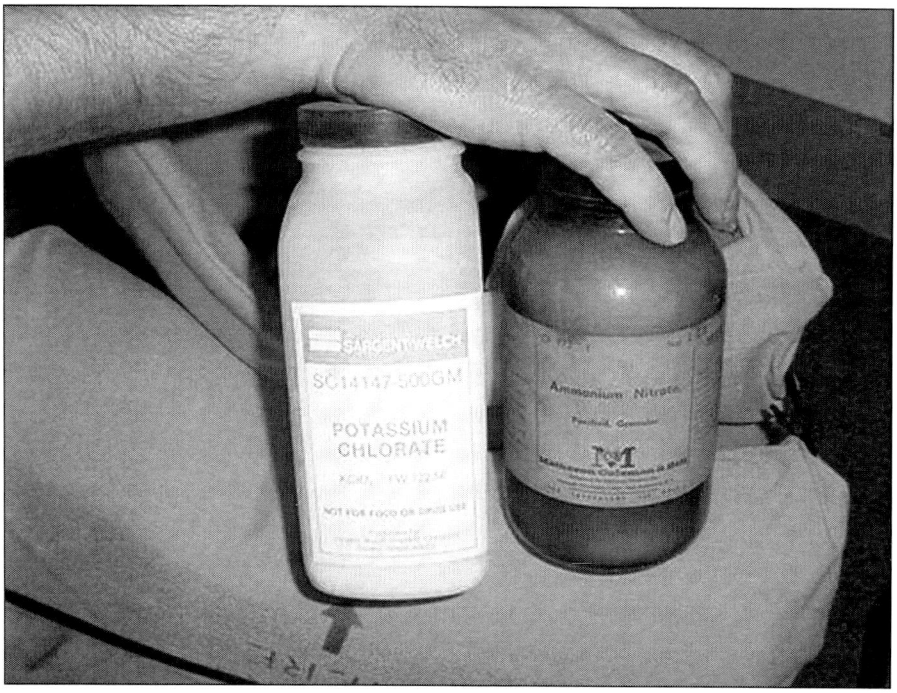

Figure 39. These explosive chemicals were removed from a high school laboratory.

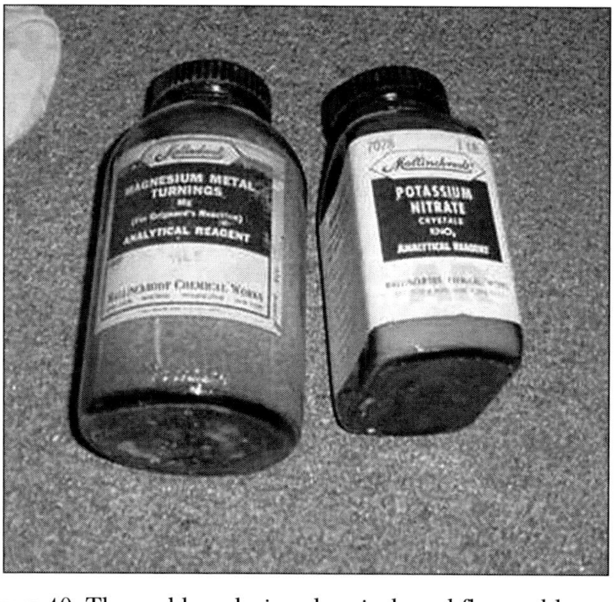

Figure 40. These old explosive chemicals and flammable metal were removed from a high school laboratory.

Figure 41. Many laboratories will have unknown chemicals with no labels in a deteriorated condition. These were found in a high school laboratory.

Many unknown liquids, solids, and reagents may be present. Only expert chemists with explosives experience along with bomb technicians should enter or handle any of the components of unknown or suspected homemade explosives. These may appear as powders or liquids. In fact, many of the same items used to manufacture methamphetamine and other illicit drugs, such as ammonium nitrate, iodine, hydrochloric acid, acetone, toluene, and sulfuric acid, are found in some homemade explosive laboratories. Remember using a methamphetamine test kit may cause an explosion if the substance tested is an explosive (see Figures 41 & 42).

Five Significant Terrorist Bombings Involving HME in the US:

- 1971 Sterling Hall University of Wisconsin (ammonium nitrate based)
- 1993 World Trade Center Bombing (urea nitrate based)

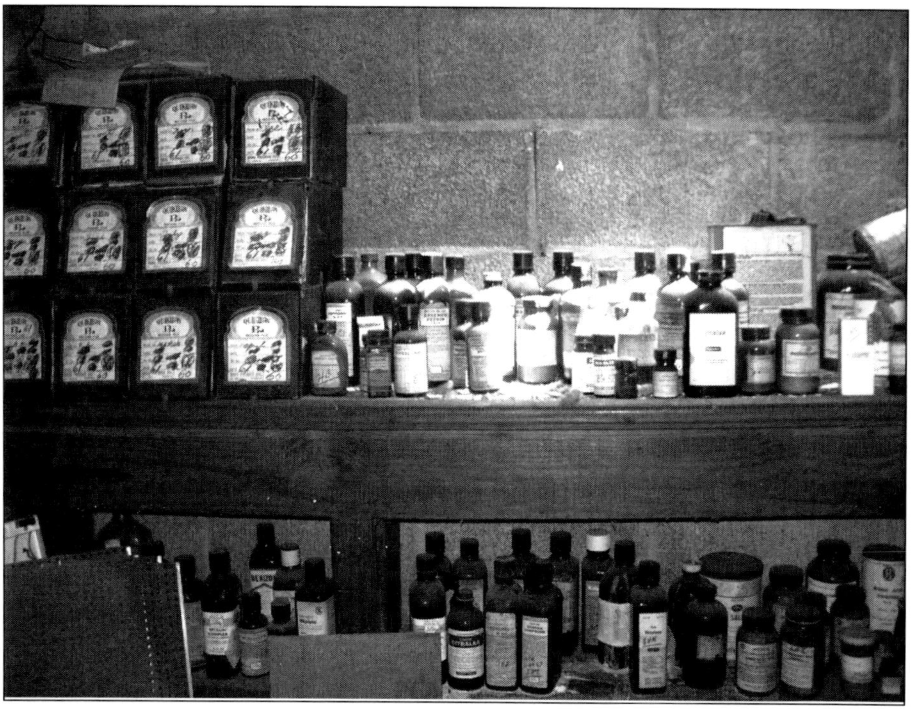

Figure 42. Bomb technicians found this abandoned business with hundreds of chemical containers. Many were unlabeled and deteriorating. This event was a major hazardous materials incident. Numerous explosive and toxic chemicals were present.

- 1996 Oklahoma City Bombing (ammonium nitrate based)
- 2003 Chiron Life Science Center California (believed to be peroxide based)
- 2003 Shaklee Corporation California (believed to be peroxide based)

Most HME mixtures will be a chlorate mixture and juveniles will create many for experimentation purposes. Some will be used for the illegal manufacture of fireworks. The peroxide based explosives are very common in experimentation as are the AN based explosives.

Indicators of a HME include the common items listed below, those with an asterisk (*) indicate such can be used to manufacture illicit drugs:

- Glassware with cooling baths, funnels, evaporation columns, filters*
- Items in a refrigerator

- An ice maker*
- Blenders, grinders, stirrers, mixers*
- Nitric acid
- Sulfuric acid*
- Hydrochloric acid*
- Fireworks
- Petroleum jelly, wax, paraffin, used as plasticizers or binders
- Potassium permanganate, used as a hypergolic ignition agent, algaecide
- Rock salt, sodium chloride*
- Protective gloves, face shields, goggles, protective clothing*
- Air purifying respirator*
- Ventilation equipment*

If peroxide explosives are manufactured, look for:

- Peroxides, 30% or stronger common used in hair bleaching or pools
- Peroxides of 2% are common in methamphetamine laboratories*
- Acetone, organic solvent and fingernail polish remover*
- Hexamine fuel tablets
- Methyl ethyl ketone, organic solvent or fingernail polish remover*
- Citric acid

If nitrogen based explosives are manufactured, look for:

- Ammonium nitrate*, potassium nitrate, instant cold packs*
- Urea, urea nitrate, fertilizer or deicer
- Potassium nitrate based fertilizer
- Toluene*, organic solvent
- Acetone, organic solvent fingernail polish remover*
- Erythritol, sweetener
- Benzene*
- Nitrobenzene, used as a sensitizer or fuel
- Ethanol, drinking alcohol*
- Iodine, may be in a tincture form or blue-gray crystals, reddish stains*
- Nitromethane or nitropropane, racing fuel, used to sensitize ammonium nitrate

- Confectioner's sugar or table sugar
- Fuel oil
- Ethylene glycol, antifreeze

If gunpowder, flash powder, or chlorate explosives are manufactured, look for:

- Potassium nitrate, sodium nitrate can be in fertilizer form, or urea nitrate
- Sodium or potassium chlorate, or perchlorates
- Charcoal or graphite
- Sulfur
- Powdered metals such as aluminum or magnesium
- Note that magnesium may be found rarely in some methamphetamine labs*
- Red phosphorus, red stains and filters with red stains*
- Match striker strips for red phosphorus*
- Match heads for potassium chlorate or air bag initiators

If chlorine based explosives are manufactured, look for:

- Ammonium hydroxide
- Anhydrous ammonia*
- Ammonium Chloride
- Electrolytic apparatus*

If thermite or thermate is manufactured, look for:

- Iron oxide, reddish powder, not be confused with red phosphorus
- Aluminum powder, gray powder
- Barium perchlorate or barium nitrate, white to off white powder
- Magnesium strip, gray metal strip
- Sulfur, yellow powder

Acetylene

A popular explosive, which is very shock sensitive and subject to explode with handling, is acetylene. Acetylene is placed in balloons, plastic bags, or other containers, and ignited. The ensuing explosion is impressive. Just a few party balloons containing acetylene contained in a vehicle will destroy the vehicle if detonated. The simple squeezing of the balloons, exposure to static electricity, or flame will cause detonation. The agitation from the air passing through a vehicle with the windows down driving at speeds under 25 mph has detonated balloons filled with acetylene contained in a plastic garbage bag. The ensuing explosion destroyed the vehicle and injured the five occupants.

Acetylene is used in welding and in some rare cases in lamps using calcium carbide. These lamps use the acetylene generated by water reacting with calcium carbide. Some hobby cannons also use this same method to generate acetylene, which is used as the fuel for the cannon.

Acetylene is so shock and pressure sensitive, it comes stored in a metal cylinder under pressure dissolved in a solvent such as acetone. The cylinders also contain a porous substrate such as silica to further disperse any compressive forces or shock the acetylene will encounter.

This gas has a wide explosive limit that some references cite as 2.5% lower explosive limit (LEL) to 100% upper explosive limit (UEL), while others show a LEL of 2% and a UEL of roughly 80%. Acetylene becomes explosive and will detonate when heated. An interesting note is that acetylene when combined with mercury, silver, chlorine compounds, cobalt, halogens, or copper forms explosive compounds, which are unstable. Acetylene may spontaneously explode if mixed with oxygen.

Airbags

Although airbags are not thought of as explosives, they are a target of thieves and do contain materials that can generate a tremendous amount of heat and nitrogen gas. Airbags are initiated during a collision when pyrotechnic devices similar to an electric match ignite sodium azide (NaN3), a primary explosive, which in turn ignites a potassium nitrate (KNO3) and sodium azide mixture. This reaction produces nitrogen gas, which fills the airbag and heat. The nitrogen gas deploys the airbag at a speed near 200 mph.

Information from a variety of sources indicates that as little as 1.5 volts DC can initiate the airbag. This could occur from static electricity. Airbags are also flame sensitive and will deploy when exposed to flame or high heat. Airbags should be handled carefully and treated as an explosive. Protect them from static electricity, flame, impact, and heat.

Chapter 4

BOOBY TRAPS

Booby traps are defined as a concealed device, which through actions of a person, causes the device to function. Although not common, booby traps are found as an active triggering device on some explosive devices. One may find marijuana "grows" or clandestine drug laboratories protected by booby traps. Open source materials show the finding of bobby traps in clandestine drug laboratories ranging from 10% to 50% depending on locale. Some agencies report almost 100% of outdoor marijuana grows are booby-trapped. Some outlaw motorcycle gangs are known for explosives and to a lesser extent for the use of booby traps. Some locales report explosives such as pipe bombs not configured as booby traps, are found within the clandestine laboratories. Many agencies report that marijuana fields are booby trapped.

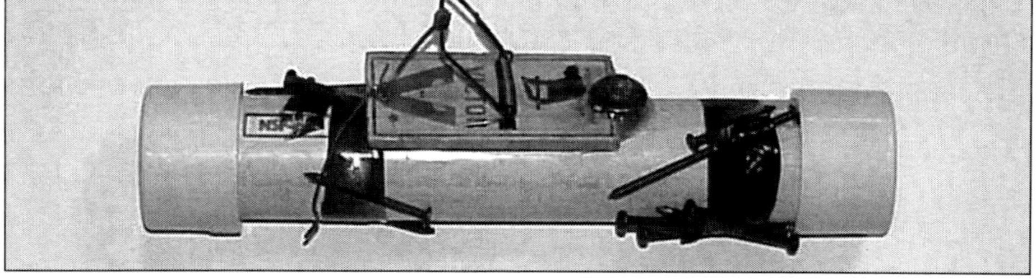

Figure 43. The recreation of a booby trap found in a marijuana field. This IED was placed behind a tree on a trail. The trip wire was in front of the tree such that as a person passed the tree and activated the trip wire, the mousetrap bale hit the BB attached to the shotgun primer detonating the smokeless powder filled PVC pipe. The tree served to direct the blast and fragments toward the person tripping the IED.

The Internet and alternative literature is replete with methods to construct both simple and sophisticated booby traps. Booby traps can be divided into categories, which include explosive, incendiary, mechanical, or chemical in terms of form. Many structures are now protected by alarm systems including passive infrared detectors and camera surveillance to warn the criminal occupants of the arrival or entry of law enforcement. A booby trap's function is usually to alert occupants, to impede entry, or to deny use of an area. Booby traps can be divided into categories by location. Primary booby traps protect a perimeter of a structure or marijuana field, while secondary booby traps are within a structure or within the perimeter of a field, and tertiary booby traps are concealed within an object (see Figure 43).

Common mechanical booby traps include pungi pits, which are a hole, covered with a material such as cloth or thin board with soil or leaves on top. When stepped upon, the foot sinks into the pit where some sharp instrument such as nails in a board impale the foot. More sophisticated traps may also use downward facing spikes to make removal of the foot difficult. Feces are sometimes applied to the nails to enhance the potential for infection of the wound. Another common form is a board covered with soil or leaves such that stepping on the board forces a rifle or shotgun shell onto a nail discharging the round. Trip wires may be used to detonate an explosive, ignite an incendiary, trigger a chemical device producing cyanide gas, or cause the release of a heavy hanging object to fall upon the person activating the tripwire (see Figure 44).

Most booby traps require some action by the person present. Some common booby traps include: tripwires hooked to a door or across a trail designed to discharge a firearm, fine monofilament line hung at eye level with small fishhooks, light bulbs filled with smokeless powder or a flammable liquid, a trip wire activating two hypergolic chemicals to mix, cyanide mixing with an acid to produce a toxic gas, or a tripwire simply attached to a small firework device to alert the occupants of a person crossing the perimeter. Some of the more sophisticated devices use electrical power to detonate an explosive when a refrigerator door is opened. On rare occasions, electronic triggers such as a passive infrared sensor may be used. The items used to detect intruders in a camping situation are suitable for such when wired to an alarm or explosives (see Figure 45).

Figure 44. This booby trap was removed by the author from the approaches to a marijuana field. (Photograph by Sergeant Gary Coleman.)

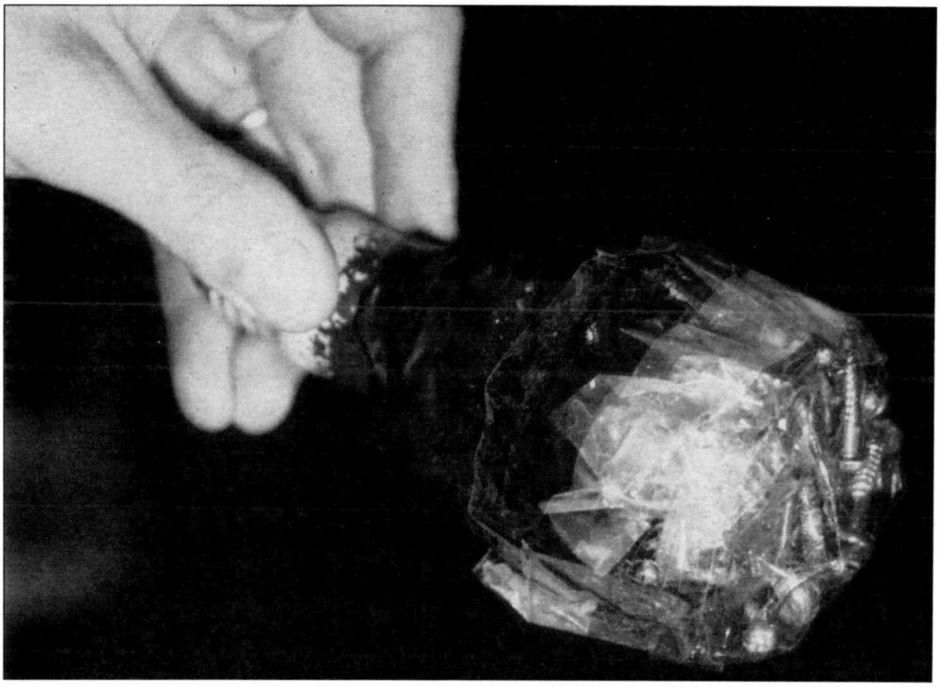

Figure 45. The author removed this smokeless powder filled light bulb from a clandestine drug laboratory. When the light was turned on, the heated filament ignited the gunpowder causing an explosion.

The use of aluminum foil coated with Armstrong's mixture produces a booby trap. The mixture is red phosphorus and potassium chlorate with a small amount of sulfur, or other strong oxidizer mixture. Even the friction of stirring the mixture has been known to detonate it. This is a primary explosive. These will likely be found in clandestine drug laboratories. Although this reddish colored aluminum foil ball appears as a non-threatening device, rough handling, kicking, or stepping on the device may detonate it.

Other common booby traps include a mouse trap with tripwire and improvised firearm pointing down a trail such that when released, the mousetrap strikes a small BB or ball bearing on a shotgun shell discharging it through an improvised barrel such as a pipe. Wooden clothes pins with a wire on each jaw separated by an insulator such as a piece of plastic attached to a tripwire, such that the tripwire pulls the insulator out and the clothespin closes completing a circuit. The circuit could detonate a blasting cap and explosive charge or sound an alarm (see Figures 46 & 47).

Military booby traps and commercial booby traps with tripwires are available from multiple sources. One marijuana grower cunningly hid a pipe bomb with nails as enhanced fragments behind a tree on a trail to his marijuana grow. When the tripwire located past the tree was triggered, the device located at the base of the tree detonated behind those on the trail. That helped direct the fragments down the trail and served to "shape" the trajectory of the fragments and blast. Some booby traps are used to attract the intruder such as currency, a valuable object. When approaching the object, the person "blunders" into a booby trap or the object itself may be booby-trapped. Be wary of overturned furniture or objects in the normal pathway. Inside windows and doors are a common point of locating tripwires and booby traps.

Tripwires are hard to see and some officers use a stick with a bright colored piece of string tied to end. This is carried in front of the officer such that contact with a tripwire is visible and felt. Some use toilet paper for this purpose or Silly String® to detect tripwires without triggering the wire. Silly String can be sprayed and if it falls to the floor or ground, no tripwire is likely present. Other officers use a handheld laser with a "slit" beam which assists in visualizing trip wires. Otherwise, the string makes the wire more easily seen. In some instances, even the weight of the tripwire detectors may cause the device to explode.

Figure 46. This booby trap guarded the front of a clandestine drug laboratory. A trip wire attached to the door initiated the explosive. A clothespin had a piece of cardboard held in the jaws of the clothespin keeping two wires separated. When the tripwire pulled the cardboard loose, the circuit was completed and the explosive detonated.

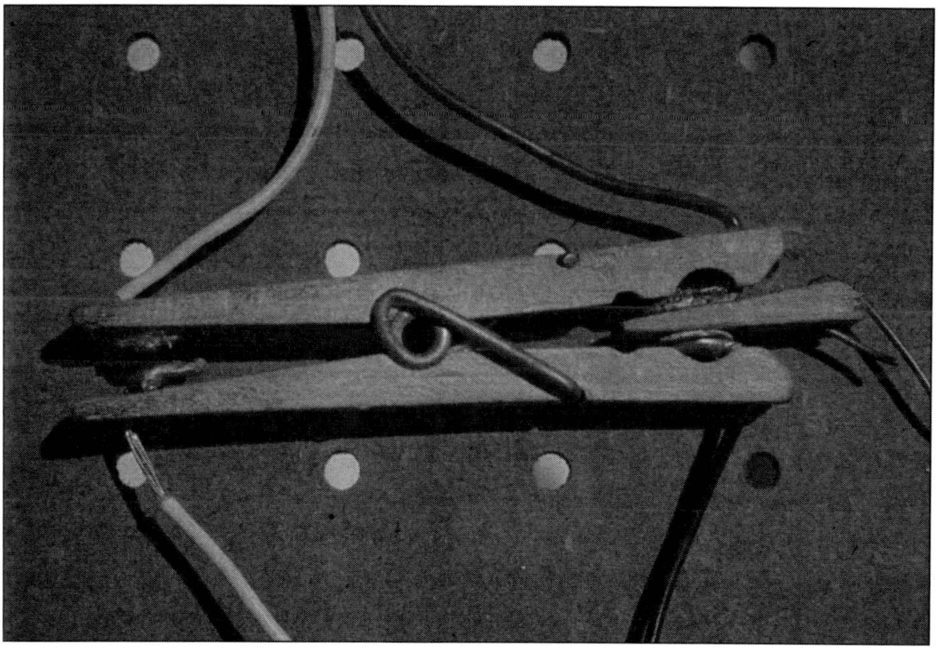

Figure 47. An example of a clothespin booby trap device.It is configured such that removal of the wood piece holding the jaws apart completes an electrical circuit and fires the charge. A tripwire would usually be attached to the wood chip in the jaws.

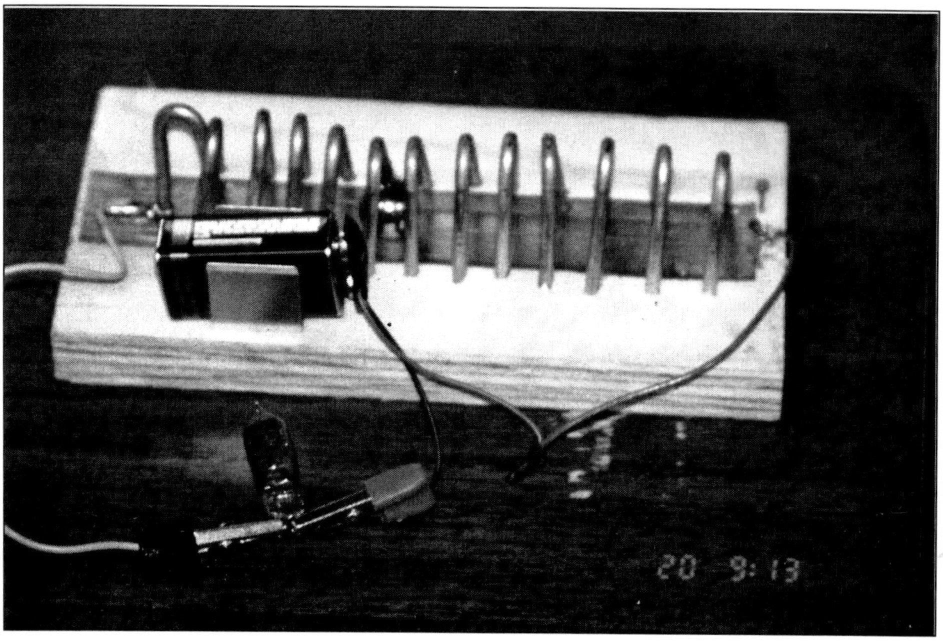

Figure 48. An example of a motion-sensitive switch for an IED. If moved, the ball bearing makes contact with metal staples completing the circuit and initiating the IED. (Photograph by Sergeant Gary Coleman.)

A classic booby trap is a package or object with an antimotion device. The classic antimotion device is a mercury switch, but as these devices are now rare, substitutes such as a ball bearing in a metal cage can be used. The key is that motion closes a circuit, which ignites a blasting cap, and in turn, the main charge (see Figure 48).

Tertiary booby traps are concealed within objects. Some of objects reported as used within the US as explosive booby traps include police flashlights, refrigerator, portable radio, cellular telephone, landline telephone, videocassette tape, candy boxes, cookie tins, packages, and even a pistol. Booby traps will likely be located upon the traveled area and approaches to the protected area. This will include trails, doorways, windows, or other entry points. Doors are not only "fatal funnels" for shooters but also for booby traps. Remember constricted areas are common location for booby traps. One should be wary of opening doors or windows, or entering through open windows or doors. Objects such as overturned furniture or a rug may conceal a bobby trap, or are designed to divert the officer's path into a booby trap. Objects, which appear "out of place," should be avoided. Rugs

Figure 49. A recreation of a device found by bomb technicians. The target was experiencing a contentious divorce and found the cookies on her front porch. She became suspicious and did not touch the package. Bomb technicians found the item had no tamper resistant sealing and a small hole was drilled into the side. The hole was for a safety to allow the device to be transported without functioning. (Device built by bomb technician Corporal John Skipper.)

may conceal "pressure applied" type triggers, or even conceal an area with no floor. If one booby trap is detected, carefully extricate oneself using the same footsteps as made on the approach. If one decides to continue forward, an open easily detected booby trap may be used to direct the path into a concealed booby trap. Outdoors, when feasible, avoid trails and travelled areas, as they will be area where booby traps are likely. Look up as trees may conceal deadfalls or other booby traps. Look for disturbed leaves or soil; look for soil from holes, which might conceal a pungi pit. Deadfalls are gravity-powered booby traps in which a trigger releases a heavy object to fall on the person by free fall, by tilt, or swinging. Most deadfalls have a payload, trigger, and load line. The trigger will usually be a tripwire attached to the anchor point of the deadfall and the line, when pulled, disengages the anchor point. This may be a simple "pin in the eye" trigger. Spikes or cutting instruments may be added to enhance the damage caused by deadfall.

Figure 50. A recreation of the IED with the lid removed. The item would detonate if the lid were removed. (Device built by bomb technician Corporal John Skipper.)

Deadfalls can be simply a bucket or other heavy object balanced on a partially open door (see Figures 49 & 50).

The key to dealing with booby traps is intelligence. One needs to be aware that such may be present. This changes the manner in which one operates and approaches a location. Bomb technicians will be needed and the approach will have to be slow and methodical. Some agencies use robots to detect and clear a path for officers. One agency uses a heavy object hand thrown with several strands of detonation cord attached and literally "blasts" an approach. This will clear tripwires, may detonate some booby traps, and will reveal many underground booby traps.

Chapter 5

TOXICITY OF EXPLOSIVES AND PLACARDING

Explosives are complex chemical compounds, which by their very nature are usually toxic and may be absorbed into the body in a variety ways. During World War II, US government sources documented that more than 400 munition workers died of exposure to TNT and more than 15,000 exhibited signs of TNT poisoning. A consideration when dealing with any explosive is knowledge of the toxic effects, methods of absorption of the explosive, and proper protective measures. Another consideration is the fact that unknown explosives must be assumed to be most toxic compound and protection must be of a maximum nature in these cases.

Explosions may be a dispersal method of chemical warfare, biological warfare, or radiological dispersion device. One scenario uses the disruption of the suspect package by bomb technicians as the dispersal method of the toxic payload. Protective clothing and respiratory protection should be considered when handling explosives in a chronic setting or explosion environment. The minimum protection should be an air-purifying respirator.

Common Toxic Materials Produced in Explosions

The combustion products during an explosion are toxic They usually contain dispersed particles of the explosive, carbon monoxide, carbon dioxide, soot, sulfur compounds, and nitrogen products. In confined spaces, these products may concentrate or linger following an explosion. The sulfur products are usually sulfur dioxide (SO^2) and rarely hydrogen sulfide (HS^2). Both are toxic. The nitrogen compounds produced include nitrogen tetroxide, nitrous oxide, nitric acid, and rarely, hydrogen cyanide (HCN). These gases tend to be acid and can cause pulmonary issues if inhaled. When feasible, explosion sites should not be entered for at least 30 minutes or if enclosed, until ventilated. Lifesaving measures usually necessitate entry, but simple decontamination of victims by removing clothes and use of skin and respiratory protection by rescuers will negate most of this type exposure. When feasible, those entering explosion sites should use level B for exterior events and it is recommended to use level A for interior operations with level C used when other levels are not feasible. Standard decontamination practices should be followed, remembering that removal of other clothing eliminates a substantial amount of surface contamination. As a minimum, those involved in such a scene should wash their hands and face as soon as feasible, doff their clothing, and shower. No eating, chewing, or drinking should occur in such a site and upon exit not until the officer has been decontaminated or showered.

Bomb technicians handling explosives on a routine basis should remember that the explosives are toxic. Chemical-resistant gloves should be worn as a minimum. Hand washing, uniform changes, and showering should occur as soon as the process is completed. No eating, drinking, or chewing should occur during the handling of explosives and in areas explosives are handled.

Specific Explosive Toxicity

TNT–this explosive is very toxic and can be absorbed through the skin, by inhalation, and gastrointestinal tract. Contact produces dermatitis; TNT attacks the liver, central nervous system, and is classified as a mutagen and carcinogen.

Nitroglycerin, EGDN–easily absorbed through skin, exposure in a chronic setting may cause sudden death syndrome, heart attack, or stroke. Acute exposure can cause a dangerous drop in blood pressure.

PETN–this explosive has a low toxicity and is not easily absorbed through the skin.

Placarding of Explosives

Explosives have several placarding categories. The signs are orange with black lettering.

Categories

1.1 Mass Explosion Hazard (primary and secondary explosives)
1.2 Projection but not mass explosion (cannon shells)
1.3 Fire, minor blast (rocket motors, fireworks)
1.4 Minor explosion hazard (blasting caps, small arms ammunition)
1.5 Blasting Agents, very insensitive
1.6 Explosives, extremely insensitive

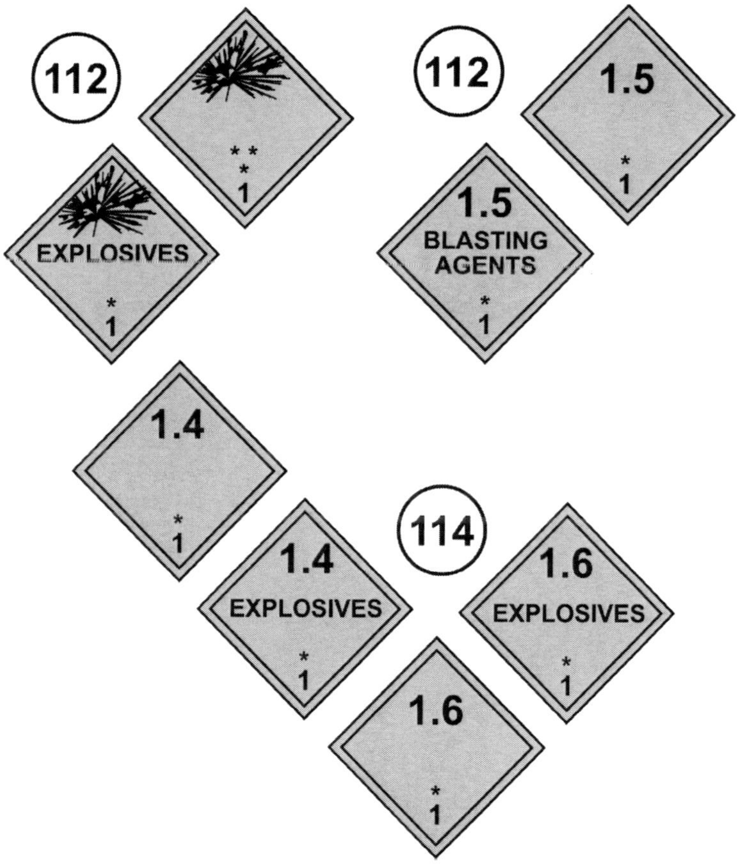

Figure 51. Placards from the NAERG demonstrating the type placards explosive shipments may bear. Containers may also have these placards present.

The placards direct one to Guide 112 in the North American Emergency Response Guide (NAERG). The important points in the guide are listed below.

Cargo Fire

- Do not fight fire when fire reaches cargo! Cargo may EXPLODE!
- Stop traffic and clear the area for at least one mile in every direction and let burn
- Do not move cargo or vehicle if cargo has been exposed to heat

Tire or Vehicle Fire

- Use plenty of water–Flood it! If water is not available, use CO2, dry chemical.
- If possible, and without risk, use unmanned hose holders or monitor nozzles
- Keep the maximum distance to prevent fire from spreading to cargo area.
- Pay special attention to tire fires as re-ignition may occur. Stand by.

Spill or Leak

- Eliminate ignition sources (no smoking, flares, sparks, or flames in the area).
- The equipment used when handling the product must be grounded.
- Do not touch or walk through spilled material.
- Do not operate radio transmitter with 330 feet of blasting caps.
- Do not clean up or dispose of materials without the guidance. The key is to evacuate and not fight fires which impinge upon explosives.

Under high heat and direct flame impingement, explosives become unpredictable in behavior and can detonate at any time. Remember the blast wave is a short-range effect, but projected fragments may travel more than one mile. Remember to stay upwind and uphill of the event when feasible.

Chapter 6

IMPROVISED EXPLOSIVE DEVICES (IEDs)

An improvised explosive device is any explosive device that is not commercially manufactured specifically for the manner in which the device is used. Only resources, the technical skills of the builder, and the mental abilities of the builder limit improvised explosive devices. In the US, most bomb squads face only single IEDs of simple construction such as a metallic pipe bomb. Bomb squads are well prepared and handle such devices easily. However, the experiences in Iraq and Afghanistan have demonstrated that use of IEDs in an offensive manner, the use of PBIEDs and VBIEDs by terrorists is feasible and not difficult to accomplish.

The US has seen few VBIEDs and few PBIEDs. However, the mere facts that these events are occurring indicate they are a viable threat. Most bomb squads are not well prepared or equipped to handle such an event. Further, the skills of bomb technicians are not well developed by this lack of situational practice and experience.

Deployment of bombs and incendiary devices is likely underreported to federal-level agencies. Therefore, the open source materials of 4,000 bombings annually is likely far higher. Open source materials indicate that many of the devices fail to function with as many 30% failing to explode or ignite. Each year, several thousand incidents occur where explosives are recovered. These may be abandoned, lost, stolen, or homemade explosives. Pipe bombs are the more common bomb and open source materials estimate that about 75% of bombs are pipe bombs. The more common form of incendiary device is a Molotov cocktail. Improvised Incendiary Devices (IIDs) also fail in many circumstances. In the typical year, according to open source materials, roughly 70 people are killed in bomb-related events. This includes several bombers killed during the manufacture of the explosives or the IED.

Open source materials estimate that at least $250 million in property damage occurs with most of the losses from fires from IIDs. Some authorities state that as many as 1,000 IEDs and more than 700 pieces of military ordnance are recovered annually. Mailboxes remain a popular target for bombs, especially bottle bombs, while vehicles and schools are common targets. Placement in buildings is usually in a public location such as a restroom, while vehicles have the device attached externally as the more common location within the interior while those placed on the exterior target machinery commonly. In schools, lockers are a common target as are trashcans. Most of the devices will employ low explosives such as flash powder or gunpowder, while IIDs use a flammable liquid such as gasoline in a frangible container. Homemade explosives are becoming more popular as explosive filler in IEDs. A few IEDs and IIDs were boobytrapped and motion sensitive while most used a burning delay while a few used electronic initiator systems. IIDs used impact to function by breaking a frangible container. In terms of explosives stolen, dynamite, and blasting agents were the more common explosive stolen with several tons stolen each year.

Open source materials also indicate that hoax device consume a substantial amount of resources. Several thousand hoax devices are found by public safety personnel in a variety of locations with suspicious objects, which are unintentional, and hoaxes composing almost 90% of the incidents. However, roughly 10% of the calls are devices assembled to resemble a live IED. Some are preceded by a bomb threat. The reason for the hoax device and bomb threat may be to harass and disrupt the operations of a business, terrorize individuals, or observe response and operations of public safety.

IED Construction

The typical IED in the US is a metallic pipe with end caps filled with the low explosive smokeless gunpowder as the explosive filler. The pipe will have a hole drilled through the end cap and use cannon fuse as a burning delay or a simple hand-delivered incendiary device

Figure 52. A PVC pipe bomb that failed to detonate. The contents were a mixture of smokeless gunpowder and flash powder from fireworks.

such as a Molotov cocktail. These are assembled easily and require little technical expertise with components available for purchase. The directions to build such are available through the Internet and alternative literature.

Metallic pipe is used as the pipe contains the deflagrating smokeless powder and when the pipe fails, generates fragments that may travel more than 300 feet. One should remember the primary method of injury and damage from pipe bombs will be the fragments generated by the explosion. Some bombers use plastic pipe, but such is not as effective as metal pipe for producing fragments and containing the expanding gases generated by the deflagrating smokeless powder filler (see Figure 52).

Some bombers will add enhanced fragmentation to the IED. This is usually nails, screws, ball bearings, or wire. These items are projected during the explosion and should not be confused with Shrapnel. Shrapnel is a specific type projectile used during World War I to inflict casualties from overhead blasts. The incorrect use of the term Shrapnel, named after its inventor, is common. The items projected from a

Figure 53. Bomb technicians prepare IEDs for post blast training. Homemade plastic explosive, detonation cord, along with commercial dynamite were used.

bomb or IED should be called fragments. Items added to the device should be classified as enhanced fragments.

Some skilled bomb makers use hardened masonry nails, which survive the explosion and do not deform as easy as conventional nails. This provides increased penetration capability. Some bombers have coated the enhanced fragments with toxic materials such as warfarin, which is an anticoagulant. However, some of these coatings do not survive the explosion process well and many of the fragments lose the coating during the explosion from the physical shock or the thermal event (see Figures 53 & 54).

On rare occasions, high explosives may be used within the metal pipe. Examples of smokeless gunpowder pipe bombs include the Centennial Park Olympic bombing. Eric Robert Rudolph, Centennial Park and abortion clinic bomber, used high explosives in pipes in his later bombs at the Sandy Springs and Birmingham abortion clinic bombings. Dynamite was the filler for these pipe bombs. In these devices, a blasting cap is required to detonate the pipe bomb. This will require a power source, usually a battery, to ignite the blasting cap. In rare circumstances, a burning fuse delay may be used with a fuse-type blasting cap.

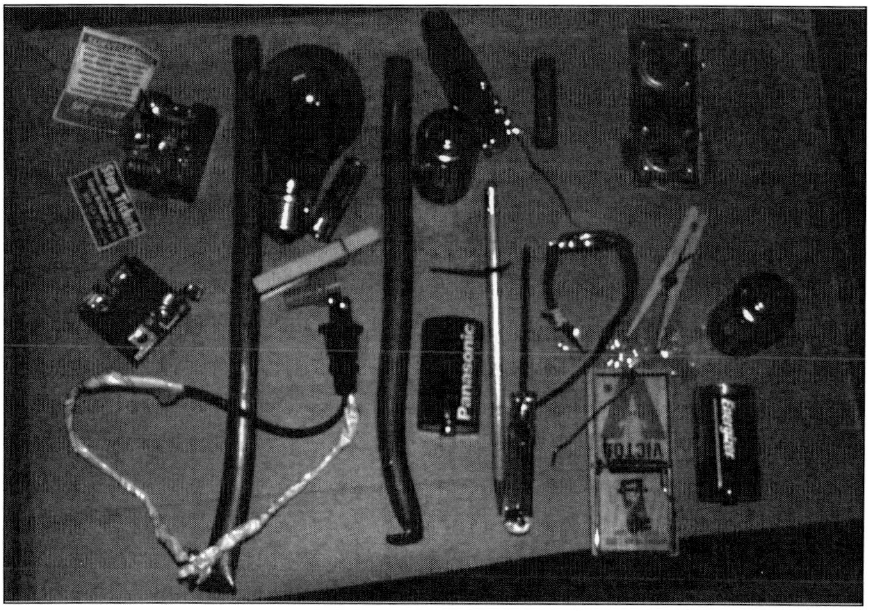

Figure 54. Unassembled components of a pipe bomb whose destination was a high school. Officers were able to intercept the bomb prior to its construction, but not before the alleged bomber killed one individual and injured two others.

Figure 55. Three sticks of dynamite are shown detonating in the front of this vehicle.

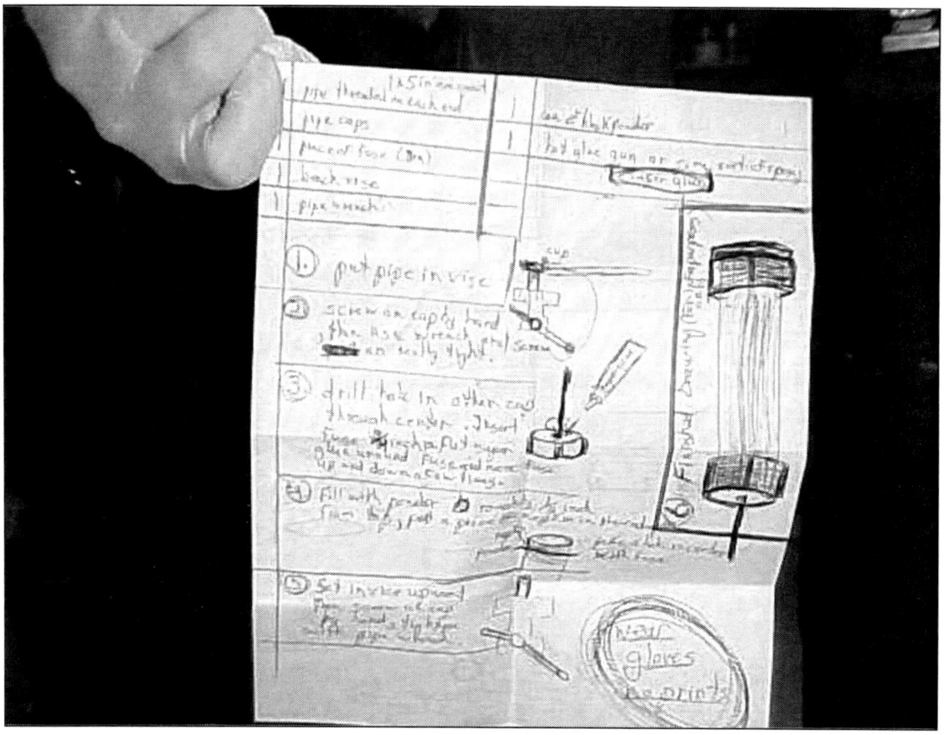

Figure 56. A bomb technician holds the directions on how to assemble a pipe bomb. The directions note for the maker to wear gloves to avoid fingerprints. The individual was attempting to direct another person in the art of pipe bomb construction.

Pipe bombs using smokeless powder use cannon or hobby fuse as the delay in many instances. However, some may use more sophisticated electric matches powered by batteries with a timer. Improvised ignition systems may include a flashlight bulb or incandescent Christmas tree light with its filament exposed to the smokeless gunpowder. An electric current, usually from a battery, will heat the filament and ignite the gunpowder. This same concept can be used when primary explosives such as TATP, HMTD, or MEKP are used as the explosive filler. Secondary explosives such as PETN or RDX require a blasting cap to detonate (see Figure 55).

Mechanical timers are usually mechanical in nature such as a wind-up alarm clock in which the moving hand is used to bring wires together completing a circuit and igniting the smokeless powder. Electronic systems using a silicon rectifier to close a circuit from a timer do occur but require a higher bomb builder skill level and additional resources. Incendiaries use a burning delay in many circumstances. This

Figure 57. Bomb-making literature that was seized from bomb makers. Their devices exploded during transport injuring the five individuals constructing the IEDs. The five bomb makers required hospital treatment. What is interesting is that all of them had ruptured eardrums. (Photograph by bomb technician Corporal John Skipper.)

may be a cannon fuse delay, burning textile wick in the case of Molotov Cocktail, or even a simple cigarette inserted in a pack of matches to serve as the time-delayed igniter.

In the past, the military used a corrosion delay device with a blasting cap called a time pencil based upon its appearance. Booby traps tend to use a percussion cap fired by a spring loaded device or pressure applied which ignites a pyrotechnic which in turn ignites a primary explosive in a blasting cap. More sophisticated devices may use a silicon rectifier holding the blasting circuit open such that when power is disconnected, the circuit closes and uses a second power source to ignite the blasting cap. This requires a high degree of skill to construct. The use of command detonated devices using radio frequency (RF) controlled devices has become more common. Simple RF devices such as remote controls to toy vehicles have been used for command detonated devices (see Figure 58).

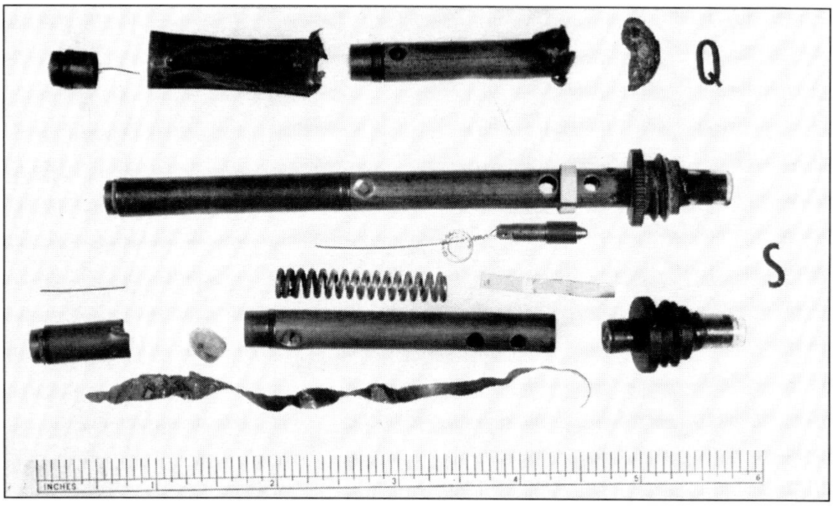

Figure 58. This is an example of a time pencil initiator used to detonate a larger charge. It first puzzled investigators until they sought military assistance to identify the device. The car belonged to a prospective future president of Cuba. Sergeant Huff identified the pieces with the help of the military at Homestead Air Force Base. Later, the author was able to dismantle three of these bombs with this information. (From Thomas Brodie, *Bombs and Bombings*, 3rd ed. Courtesy of Charles C Thomas, Publisher, Ltd.)

Container Performance and Fragment Production

Formulas exist to predict the performance of metallic containers subject to explosive forces. The geometry of the charge and the container along with the material the container is constructed of factor into the performance and fragments produced. One formula used to predict container performance and fragment propagation is the Gurney Model.

Gurney Model Container Performance

$$\frac{V}{\sqrt{2E}} = \left(\frac{M}{C} + \frac{1}{2}\right)^{-1/2}$$ for cylinders

$$\frac{V}{\sqrt{2E}} = \left(\frac{M}{C} + \frac{3}{5}\right)^{-1/2}$$ spherical charges

The Gurney equations use and connect the following quantities:

C–Mass of the explosive charge
M–Mass of the accelerated metallic materials
V or V_m–Velocity of accelerated fragments following detonation
N–Mass of the metal on the opposite side of the explosive charge
$\sqrt{2E}$ Gurney Constant for a particular explosive expressed in units of velocity

This model does not always predict the container performance and fragment formation of brittle metals such as cast iron containers. In this circumstance, the velocities will differ by as much as 20% from the Gurney predictions.

This model does predict fragment size well. A one-inch diameter steel pipe bomb 12 inches long with a 0.25-inch wall thickness will produce a range of fragments if the filler used is smokeless powder

Projected fragment sizes are:

- Five-gram fragments or large should compose 6% of the total
- One- to 2.5-gram fragments should compose 32% of the total
- Fragments less than one gram should compose 62% of the total

The fragments with more mass travel further, but the fragments with lesser masses achieve a higher velocity. Containers fail by the gas overpressure produced in low explosives and by the shock wave followed by gases in high explosives. The shock wave in high explosives starts micro-cracking which proceeds but is bypassed by the shock wave at two to five times the micro-cracking speed. The failures of the material are usually four to six times the diameter of the charge.

Two mechanisms produce structural failure are gas pressure and flexural rupture. The cracking process in longitudinal failures along the long axis of the container is called radial cracking, and is the result of the shock wave. In pressure vessels, this is called hoop stress and causes rapid container wall pressure. The gases put the container in tension because of the hoop stress exacerbating the cracking process. The gases exert pressure perpendicular to the axis of the container and the container fragments into wedge-shaped fragments. This is called flexural rupture and allows the fragments to be propelled away from the container at perpendicular angles. This same process causes spall-

Figure 59. Note the small and thinned fragments. These are characteristic of high explosives. This was a pipe bomb using dynamite.

ing, ejection of material from a container or wall, which is in compression. The reflection of the pressure wave on the opposite side followed by rarefaction causes the ejection of a fragment of the wall or vessel side.

Based upon this formula, one can expect low explosives to produce larger fragments with more mass. These fragments will have sharp edges and will have distinct linear edges near 90°. High explosives produce smaller fragments with "thinned" fragments with edges with small angles. The transonic gases rushing by the fragments during the explosions and the rapid container failure produced by the transonic shock wave produce the thinning of the metal fragments (see Figure 59).

Bombs produce two categories of fragments. Primary fragments come from the container itself, whereas secondary fragments come from the materials pushed by the blast wave and gases. Some authorities describe tertiary fragments as distant objects shattered by the explosion, which produce fragments, which fall. An example would be the glass from windows broken distant to the explosion (see Figure 60).

Figure 60. These fragments came from a small pipe bomb using smokeless gunpowder. Note the large fragments with large perpendicular edges on the fragments.

To detonate, an IED must have a firing train. The firing train may be as simple as a burning cannon fuse or as complicated as a microchip timer using a collapsing circuit with a silicon rectifier. Burning fuses use a spit of flame either to ignite a primary explosive within the IED or to ignite a primary explosive within a blasting cap. Incendiary devices use the flame to ignite the flammable contents of the device. Blasting caps have a small amount of a primary explosive which, when ignited detonates a larger base charge of PETN or RDX. This base charge detonates the device. Blasting caps may be electrically initiated or initiated by percussion as in nonelectric (Nonel) blasting caps.

Pipe Bombs

The more common device used in bombings in the US is a metallic pipe bomb. The usual filler is smokeless gunpowder. Many use hobby fuse, also known as cannon fuse, as the delay. Since this type bomb is a deflagration rather than detonation, only a spark from the burning fuse is needed. More sophisticated pipe bombs may use a mechanical timer such as an alarm clock with a battery attached to a filament from a flashlight bulb to ignite the smokeless gunpowder.

Initiation Methods

BAROMETRIC–This type initiator detonates with air pressure change and may use a balloon with its expansion during ascent bringing wires into contact or a barometer may be used. This IED would target an aircraft.

COLLAPSING CIRCUIT–This is a circuit holding the initiator open via battery or electrical power. The disruption of the power supply or the circuit closes the firing circuit, which has an independent power supply and detonates the IED. This type of IED is rare and is designed to defeat render safe procedures. Some IEDs use power stored within capacitors to power the blasting cap or other detonator. This means even disruption of the power sources may fail to stop the explosion.

COMMAND–This type IED is used in the Iraq–Afghanistan wars with regularity. This requires a human, usually with a radio frequency device, to initiate the IED at an appropriate time. Any remote device from remote-controlled toys, garage door openers, cellular telephones, or any other wireless device can be used. Eric Robert Rudolph is alleged to have used this type device to detonate the bomb targeting the abortion clinic in Birmingham, Alabama when Police Officer Sanderson approached the device.

LIGHT–This type of IED detonates on exposure to using a photovoltaic cell closing a circuit with a silicon-controlled rectifier wired to a blasting cap. This type IED is designed to function when a person opens the container. These initiation systems can be bought marketed as a device to let a person know their mail has arrived at a curbside mailbox.

PENETRATION–This IED detonates when a metal device such as a knife blade penetrates package. The blade completes a circuit between two sheets of metal foil or two wires embedded in the structure of the container. This is an antitamper feature.

MOVEMENT–This type IED may use a metal ball bearing which with movement completes a circuit, a mercury thermostat switch, or bare wire loops which close when the container is opened or moved. This is both a booby-trap and antitamper feature. This would also prevent moving the IED from its location.

OPENING–This is usually a booby-trap targeting a specific person as the device is human operated. This may be a pressure release with a mousetrap bale closing bringing wires to close a circuit when the package is opened. An even simpler method is use of a string attached to a piece of cardboard inserted between the jaws of a clothespin with wires on both jaws. This completes the firing circuit once the cardboard insulator is removed. Another method is opening the package and pulling a bare wire loop together closing the firing circuit. In rare circumstances, a friction strip may be used to drag strike anywhere matches to ignite as the package is opened.

PRESSURE APPLIED OR PRESSURE RELEASED–This type IED is initiated by closing contacts because of weight applied or removed. This may be simple can lids with the tops separated with an insulator such as newspaper and the lids with "spears" created by a manual metal can opener. The pressure of stepping on the device punches the metal through the newspaper closing the initiation circuit. Some devices use a BB soldered to the primer of a shell placed in a container such that stepping on the IED forces the round down onto a hard surface with the BB serving as a firing pin. An alternative is to use a nail in lieu of a BB. This type device may be found protecting marijuana grow sites. Another method is a trip wire which pulls a pin from a gravity feed device striking a primer or releasing a mousetrap or insulator held in the jaws of a clothespin.

TIME–This IED detonates at a specific time by mechanical clock, electronic timer, or general time using chemical means (military time pencil) or even melting ice. The mechanical clock is the simpler of the devices but more IEDs are seen now with electronic timers using a silicon-controlled rectifier to close the firing circuit. Ice placed in the jaws of a clothespin holding two wires apart is another type of simple timer, but the time of detonation is not precise.

TRIP WIRE–This initiator when pulled, released, cut, or otherwise moved allows the IED to function. This may release the bale of a mousetrap pull. This either releases a pressure release device (mouse trap, clothes pin contacts), or pulls bare wires together. This is usually a pressure release device and is fairly common. Used in booby traps extensively around marijuana field and CDLs.

Pipe bombs using high explosives are not common. However, one technique to convert a low explosive such as smokeless gunpowder into a high explosive upon detonation is to use a blasting cap to initi-

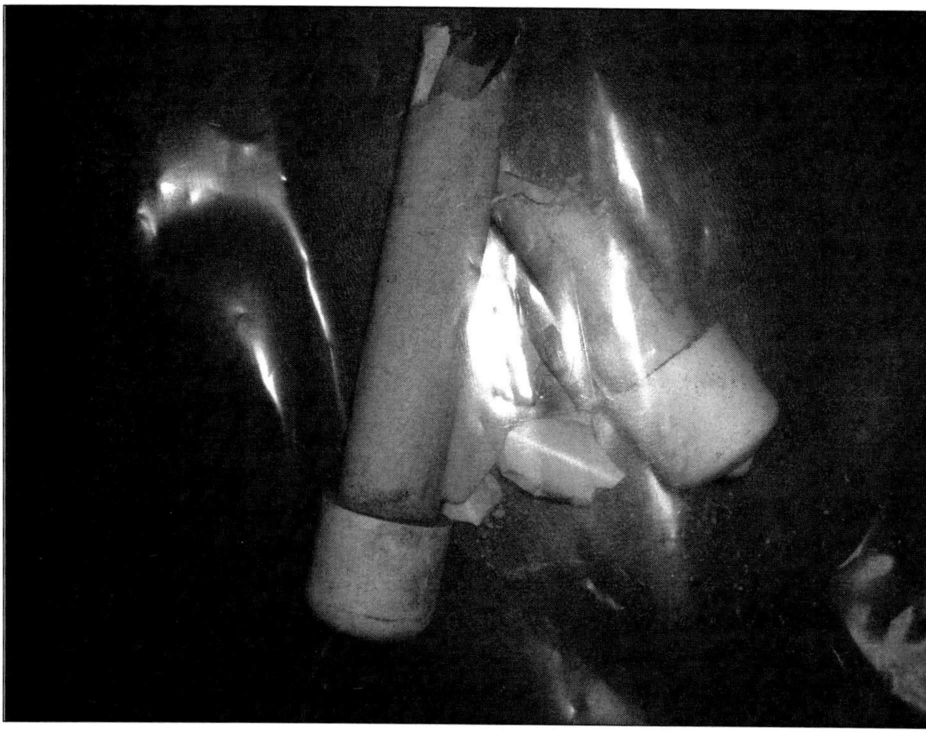

Figure 61. This pipe bomb detonated using flash powder. Note the entire pipe failed to fragment. This was a low order explosion.

ate the smokeless gunpowder. This added energy causes the smokeless gunpowder to detonate rather than deflagrate.

In many instances, pipe bombs will have a hole drilled in one end cap to facilitate a fuse of other wire to reach the inside of the pipe bomb. One serious issue with pipe bombs is that during loading with black powder, homemade explosives, or smokeless gunpowder, flakes of the material may be caught in the pipe threads. When the end cap is replaced, the friction from screwing the end cap on may ignite the particles causing premature functioning of the device. Some bombers have resorted to coating the threads of the pipe bomb with petroleum jelly to avoid this problem.

Some individuals may weld plates onto the pipe bomb as the end caps are a weak point in the device. Some bombers use pressure containers such as LPG bottles or other pressurized gas containers. The longer the device contains the gases of deflagration, the more gas is generated and the more violent the explosion.

A few pipe bombs will use plastic pipe. The obvious problem with a low explosive is the plastic pipe is the failure pressure is far lower than a metal pipe. The plastic produces fragments, which have a short travel distance.

On some occasions, the filler may be improvised mixtures such as black powder, flash powder, or potassium chlorate with metal flakes. The key is one may not know what the filler of the pipe bomb consists of until laboratory testing (see Figure 61).

Mail Bombs and Suspicious Packages

Prior to the regulation of the weight of packages, letter and package bombs were common. The weight limitation and requirement of in-person mailing have decreased the number of mail bombs. A key factor to remember is that letter bombs are still occurring as are package bombs such as those used in the Boston Marathon Bombing. These items can be anonymous, allow an attack from a distance, and via mail delivery of a package into a secure area (see Figures 62 & 63).

The UNABOMBER used the mail for many of his bombs. The factors one must consider are listed below. Key factors are the letter or package was not expected by the recipient, or the sender did not transmit the item. Other factors are the package appears out of place or the package "looks suspicious." Remember the difference between "out of place" and suspicious packages.

Other Factors:

- The package is not delivered in the normal manner
- The package is not delivered by the normal person
- The delivery method is not known
- Handwritten labels or grammatical errors
- Spelling errors, titles incorrect
- Mailed from address does not match postmark
- A known threat group named in return address
- Too much postage, stamps, or metered items
- If an envelope, thicker than a pencil
- If an envelope, weighs more than two ounces
- The envelope is stiff or bulky
- The envelope feels like it contains a powder or liquid

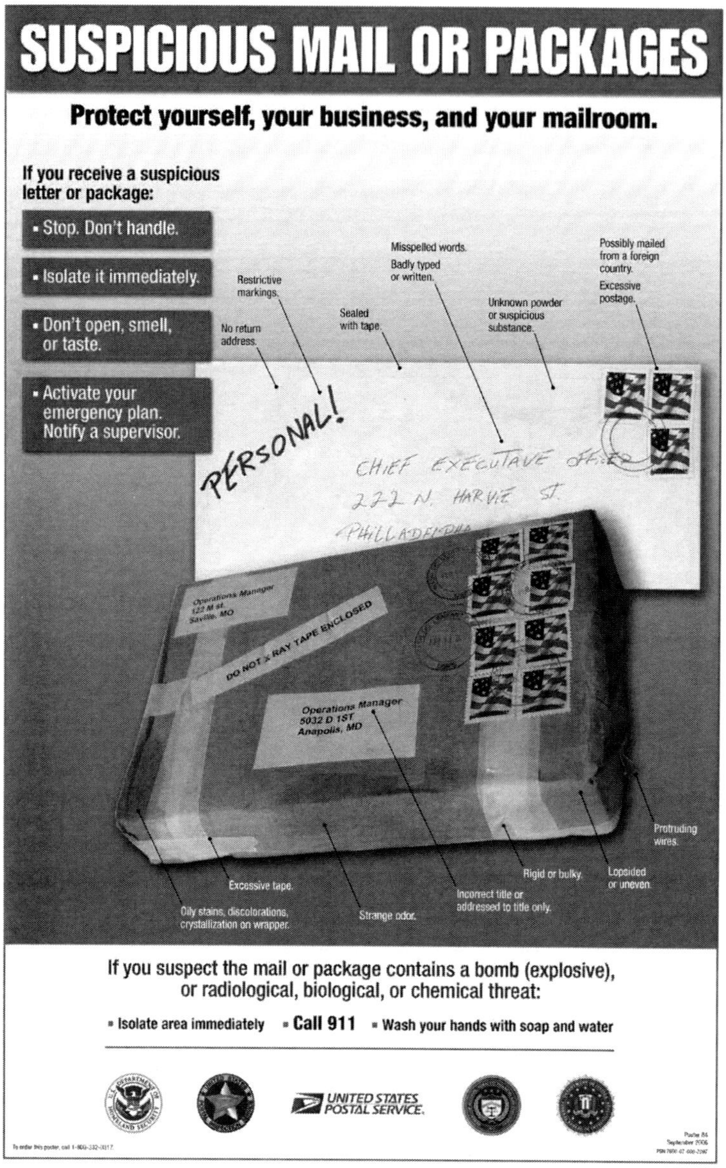

Figure 62. Suspicious mail poster.

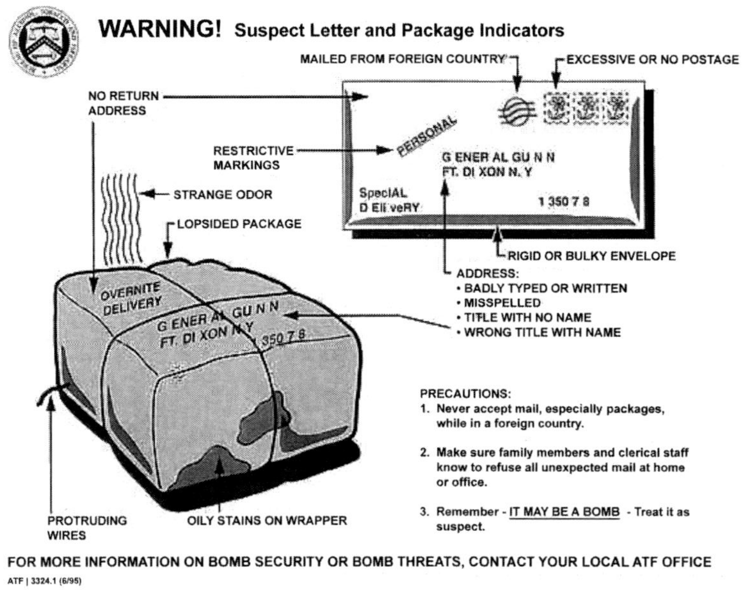

Figure 63. Another example of a suspicous mail poster.

- The envelope or package is unbalanced
- Stains or leaks, sloshing
- The delivery option is restricted, "Addressee Only, Personal"
- Item just looks "odd" or suspicious
- Package has twine, wires, metal foil, excessive tape
- The item was not professionally typed, wrapped, or prepared
- If an out of place item, the owner cannot be located.

Suspicious Packages

Out of place means the item does not appear to belong at its location while suspicious means the item meets one more of the bomb criteria listed above. If the item is suspicious, isolate it, notify the chain of command, and begin evacuations. Only one indicator may be present, but if one believes the package to be suspect, act on that information.

Suspicious packages and suspicious objects are common and one must react to the event. In some instances, the bomber may be testing security procedures and police response with suspicious packages to

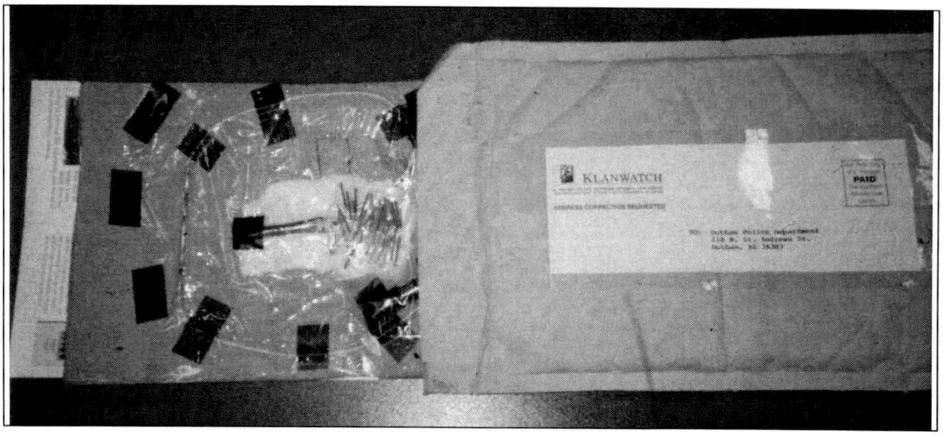

Figure 64. This letter bomb used a magnetic reed switch to initiate the IED which was a homemade plastic explosive with a homemade blasting cap and added fragments in the form of finishing nails. This is a reconstruction of the IED.

gauge the best tactics to defeat the tactics used. Another reason suspicious packages may be used is to "desensitize" and fatigue responders. This may cause a loss of vigilance and cause procedures not to be followed. Packages sent to those visible in the organization, those with media attention, or those involved in controversy should have a higher level of suspicion if an unusual package or letter is received.

Any suspect item should be isolated, the area around the item or package evacuated to at least 50 feet. Envelopes should have at least the adjacent areas evacuated. Packages, briefcases, or backpacks should have the floor or area within 300 feet evacuated as a minimum including any floors above or below the suspect item. Remember that letter or package bombs can produce a substantial amount of forensic evidence and reduce handling not only for safety purposes but also to protect any forensic data present (see Figure 64).

Envelopes or Packages with Powdery Substances

Envelopes and packages containing powders have been received in the US and most are hoaxes. However, this does not rule out a harmful substance such as bacillus anthracis or anthrax, a potent bio-toxin ricin, or radioactive material, or another harmful substance.

If the envelope meets the suspect criteria, isolate it, evacuate the immediate area, and wash hands with soap and water. Suspect envelopes should not be handled, moved, shaken, or manipulated any more than is necessary to isolate them. Close windows and doors. If feasible, close windows and doors, turn off the HVAC or other ventilation system, and turn off any fans present.

If the package or envelope has been opened and white or off-white to brown colored powder is visible, and a written threat is present, this event has must be treated as a credible event.

The following steps should be taken:

- Do not move the package or touch the powder.
- Cover the package or envelop with any available items.
- Have those not exposed evacuate the area and notify security and police.
- Notify Postal Inspectors if the item came by US Mail.
- Isolate the area.
- Close doors and windows, turn off the HVAC including any fans.
- Evacuate adjacent areas.
- Those exposed should exit the room but stay in an isolated area.
- If feasible wash hands with soap and water without leaving.
- If powder is present on clothing, remove the clothing avoiding exposure.
- Leave the clothing and evacuate to an adjacent area.
- Alternatively some agencies want the clothing bagged in a sealable bag.
- If a shower is available in this area, shower with soap and water.
- During showering, pay particular attention to hair, face, and hands.
- Environmental testing on site can be accomplished by most bomb squads.
- Hazardous materials units can also perform the testing.
- The public health agency will usually make decisions regarding offsite testing.
- This input may be needed to use a state level WMD-rated laboratory.
- Public health will make medical treatment recommendations.
- The police or FBI will make decisions regarding evidence collection.
- The main effort should be directed at contamination control.

- Decontamination of those exposed is important.
- Identification of the substance as toxic is crucial.

Threat items which have been sent via mail or a package delivery service include bombs, incendiaries, anthrax, ricin, cyanide, poisons such as warfarin, radioactive material, and other non-toxic hoax materials. Common hoax materials, which are usually "bright white," are flour, salt, sugar, and confectioners' sugar. However, one should not presume that a "bright white" powder is in fact a hoax.

VBIED

A vehicle bomb is a vehicle carrying an IED. One usually considers this a threat outside the US. However, several significant VBIED events have occurred in the US.

- September 16, 1919, a horse-drawn carriage with roughly 100 pounds of dynamite and iron pieces for enhanced fragmentation kills 40 and injures 300 in New York City Wall Street area.
- May 18, 1927 Bath School bombing, the bomber used a VBIED with more than 50 pounds of dynamite to kill several of the individuals near the original explosion site with a death toll of 45.
- August 24, 1970 University of Wisconsin in Madison attacked with a stolen van with an estimated 2,000 pounds of ANFO and one person was killed.
- September 1988 an IRS office is bombed in California using bombs in a stolen vehicle parked in the building's garage with minimal damage.
- February 1990 an attempted bombing of IRS office in California with a powerful VBIED which fails to function, the device contains 2,000 pounds of an alleged ANFO mixture in a stolen truck outside the IRS office.
- February 26, 1993 a large bomb contained within a rental van is detonated under the World Trade Center in its underground parking deck killing six and injuring about 1,000, with the bomb consisting of roughly 1,300 pounds of urea nitrate.
- April 1, 1993 a vehicle containing mortars fired 13 pipe bombs at an IRS office in California, nine exploded and several malfunctioned.

- April 19, 1995 a truck packed with about 4,000 pounds of nitromethane-sensitized ammonium nitrate (AN) destroyed the Murrah building killing 169.
- May 1, 2010 a primitive bomb consisting of roughly 250 pounds of urea nitrate mixed with sugar malfunctions in Times Square in New York City.

Substantial VBIED Incidents Outside the US:

- April 18, 1983 the Beirut Embassy in Lebanon is destroyed by a suicide VBIED. Some 63 are killed including 17 Americans.
- October 23, 1983 Shiite suicide bombers in Beirut Lebanon drive a truck loaded with 20,000 pounds of explosives into the US military barracks and detonate a bomb while a second bomb targeted French Paratroopers. Several defensive measures are defeated by the suicide bombers.
- December 12, 1983 the US Embassy in Kuwait City, Kuwait is attacked by Shiite suicide truck bombers. The attacks result in five deaths and around 80 injured.
- September 20, 1984 Beirut Lebanon a truck bomb exploded near the US Embassy Annex killing 24 persons. Reports vary as to the type and quantity of explosives with some reports citing up to 1,000 pounds of explosives while others cite lesser amounts.
- August 7, 1998 two suicide VBIEDs are used to attack US embassies in Nairobi Kenya and Dar es Salaam Tanzania with each reportedly about 1,800 pounds of TNT, and killing approximately 210 in Nairobi and 11 Tanzania while injuring more than 4,600.
- September 16, 2008 the US Embassy in Yemen is attacked with a VBIED, which kills 16.

VBIEDs present the ease of loading large amounts of explosives into a single platform and when driven by a suicide bomber, these vehicles are hard to stop. The ability of suicide bombers operating VBIEDs to penetrate security is demonstrated by the bombings in Nairobi and Beirut. The other method is by stealth in which a vehicle is parked in a vulnerable location and the bomber leaves, with the bombs on a time delay. Some bombers have used concrete and sand bags to deflect the outflow of gases to concrete the energy toward a portion of the structure.

In some instances, the VBIED may not be timed for an explosion but may use a remote detonator using a radio frequency device. The threat of IEDs has become common and a consideration by law enforcement and security personnel. What has not entered the mindset is the use of VBIEDs since such has been uncommon in the US.

Data from the US Department of Transportation Bureau of Transportation statistics says more than 6,000,000 trucks are in service in the US. According to data from the FBI and BATFE, the typical passenger car can be loaded with 500 to 1,000 pound of explosives. A large truck could carry nearly 60,000 pounds of explosives and smaller box type trucks, like those used in the Oklahoma City and World Trade Center bombings, can carry about 4,000 pounds of explosives.

The use of a truck borne VBIED can be considered a large vehicle borne improvised explosive device or LVBIED. The vehicles used in the World Trade Center, University of Wisconsin, and Oklahoma City bombings would meet these criteria. Another issue relating to VBIEDs is the use of an IED in combination with a hazardous material to enhance the lethality of the incident. Tankers carrying propane or a toxic gas such as chlorine or anhydrous ammonia could increase the number of casualties if the product were released by an IED.

VBIEDs can be used against both hard and soft targets as is illustrated by the US Marine bombing in Beirut where the suicide bomber penetrated several layers of security. The World Trade Center is an example of a soft target in which a LVBIED was effective. Potential targets, which would cause a large number of casualties and capture national attention, are schools, hospitals, stadiums, or government buildings.

Law enforcement and security personnel must be alert for vehicles "out of place." A truck off a truck route, parked near a sensitive location, should be investigated. Trucks with hazardous materials placards out of truck routes demand special attention. Officers who have been certified as DOT inspectors are a valuable resource when trucks are involved. These officers are familiar with federal and state regulations along with the construction and use of such vehicles.

Trucks parked in an unusual location should be investigated. The LVIEDs used at the World Trade Center and Oklahoma City bombings, along with the attempted bombing of the Times Square in New York City, used burning delays. This may fill the interior of the vehicle with gray smoke. This smoke is what alerted citizens to the vehicle in Times Square. The only indication the officer may have of a parked VBIED may be a vehicle cabin filled with gray smoke. The officer then must make a decision to try to intervene or evacuate.

The only time some VBIEDs could be detected and mitigated could occur during transit to the target. If during a stop of a suspicious vehicle, especially a regulated truck, the officer should be cognizant of driver nervousness, unfamiliarity with the vehicle, procedures, or documentation.

Just as the Marine Barracks in Beirut were barricaded, some locales in the US have barricades to resist the intrusion of a vehicle. One issue, which must be clear, is the rules of engagement for vehicles attempting to breach a barrier. Standoff distance to mitigate blast and fragment damage is critical. Gunfire is not an effective tool to stop a vehicle unless large caliber weapons such as a .50 caliber machine gun are used. Even then, the interaction of the bullets and the vehicle may cause a detonation or the driver may detonate the device.

Physical barriers such as concrete or water-filled barricades, bollards, or similar devices are essential. The longer the standoff distance, the better protected the structure will be. Some recommendations for structures are 300 feet. This is not easily achieved and may not be feasible in many urban areas. However, every foot of standoff distance is critical. Another critical consideration is what actions should be taken if a suspicious vehicle is detected. Obviously, the index of suspicion will determine the steps to be taken. A well-written policy is needed. A legally identified vehicle parked in a low risk area will likely not provoke the response of a vehicle placarded with hazardous materials illegally parked in a sensitive area.

The first steps are to determine if the vehicle is expected and to locate the driver. The assessment of the driver's behavior and documentation should provide the officer direction. A search of the vehicle, explosive vapor, or swipe testing can be undertaken, or even a canine sniff in low-risk circumstances. Officers may have to take immediate actions such as isolation and evacuation in high-risk circumstances in which a driver cannot be located, suspicious objects such as wiring are observed or smoke in the cabin of the vehicle is present. Officers may need to flee and place a solid object between themselves and the vehicle while avoiding overhead glass.

These incidents typically do not fall within either extreme and the officer is required to make a judgment call based upon sketchy and even conflicting data. The agency policies should address the following areas:

- Vehicles with a low to moderate risk parked in low- and high-risk areas
- The inability to locate a driver in vehicles with low, medium, and high risks associated with where the vehicle is parked and its appearance
- What parts of the vehicle should be examined or searched
- What steps to take if a driver is located, what documentation to examine, and a policy on searches
- Explosive canines *will not be deployed in high-risk* scenarios to protect the handler and canine
- When electronic explosive vapor detectors should be used
- What time constraints are available for officers to act
- When should evacuations occur or should the vehicle be moved
- When should the bomb squad respond
- The actions to take when a presumptive VBIED is present

When the officer makes a traffic stop, this provides a chance for an interview and inspection of documents. Some DOT-trained officers may even be allowed by regulation to inspect a cargo. When making a traffic stop, select a good parking location which is safe from traffic. Officers should assess the documentation and driver nervousness carefully. Unusual license plates, a driver's license, which does not allow the driver to operate such a vehicle, should lead to a high index of suspicion.

Determine if the vehicle is transporting any hazardous materials. The driver should be questioned carefully about the cargo, where the cargo came from, and where the cargo is to be delivered. The driver should be familiar with any aspect of the vehicle, cargo, route, and documentation.

Some indicators of out-of-the-ordinary issues include:

- Trucks with hazardous materials such as explosives, chlorine, anhydrous ammonia, out of truck routes, and in densely populated areas
- Trucks with hazardous material placards parked near malls, hospitals, government facilities, schools, or other heavily occupied areas

- Trucks with toxic gases in residential or business districts not near truck routes
- Gasoline tankers more than 50 miles from their point of origin as most are short-run haulers
- Rental trucks in out of the ordinary locations or a rental truck with hazardous materials placards
- Drivers without commercial driver's license or carrying hazardous materials without a hazardous materials endorsement
- Any vehicle with explosive placards which is unattended

A serious situation arises when a vehicle meets some of the criteria, but a driver cannot be located. The officer may have to proceed based upon his or her assessment and intuition. A more critical situation is in the above circumstance with a positive canine or explosive sensor alert. One must presume a VBIED is present and react as such.

Procedures must allow officers to take unilateral actions without consulting a supervisor because of the time-limited nature of these incidents. Evacuation of the immediate area is imperative. Immediate response of the bomb squad is essential as their advice and equipment may allow a better assessment and remote assessment via robot. They may also calculate an evacuation area. The only certain tactic is prevention as VBIED, when driven by a suicide bomber, have defeated many countermeasures.

Shaped Charges and Explosively Formed Projectiles

Explosives can interact in a unique manner with explosives and other materials to form jets of molten metal, water, or other substances to produce work. The common configuration for metal jets is metal placed in wedge or conical shape backed by explosives. Copper, mild steel, aluminum, and on rare occasions, glass are used to form the jet. Water is also used as a jet for penetrating objects or cutting. The explosion forces the metal or other substance, into a jet, which has a velocity approaching the velocity of detonation.

Velocities of 15,000 fps are not uncommon and some tips of the jets may approach 30,000 fps depending upon the explosive used. Metals melt, but the heat has little to do with the materials penetrating ability. The jets form a cylinder shape with a thin, very fast jet in front with

Figure 65. Shaped charges come in a variety of configurations.

a larger "slug" of the material following the thin tip of the jet (see Figures 65 & 66).

This jet uses it kinetic energy to push aside and push through materials. As the jet impacts the substrate to be penetrated, it forms a crater as the thin tip of the jet begins to push aside the matrix of the substrate. As the main body of the jet impacts the substrate, penetration continues at a higher rate. Substantial penetration of steel, ceramic armor and concrete is common. These charges are used in the industrial cutting processes and for demolition (see Figures 67 & 68).

They are designed in conical, line, and other unique shapes. Water is used in many industrial and law enforcement settings in which penetration to a specific depth is needed with the reduction of unintentional damage to materials behind the item to be penetrated or cut. Explosive entry uses this concept to cut doors, walls, glass, or doors without generating dangerous long-range fragments. These jets can be directed with precision and formulas exist to allow precision cutting or penetration.

Figure 66. Bomb technicians responded to a call when children found this 81mm mortar round. The round is in pristine condition. Bomb technicians suspected the round was not live but used shaped charges to destroy the round after consulting with military EOD.

The author has frequently used 48-gram PETN powered conical charges with a copper slug to penetrate several inches of steel armor and the jet continued more than six feet into hard compacted clay. This same technique using IV fluid bags and specially designed water containers is used to penetrate doors and windows for breaching.

The author has used this technique to penetrate wooden doors without shattering glass windows in the door during a tactical breach. Water resists compression and transmits the kinetic energy from the explosive directly into the material creating a high-speed jet, which penetrates and cuts the material. The water jet dissipates rapidly and does not present a "fragment" hazard. However, one must consider that some fragments are generated by the process and can present a hazard on the opposite side of the material cut. Back blast is also generated and can affect persons standing behind the breaching or cutting charge.

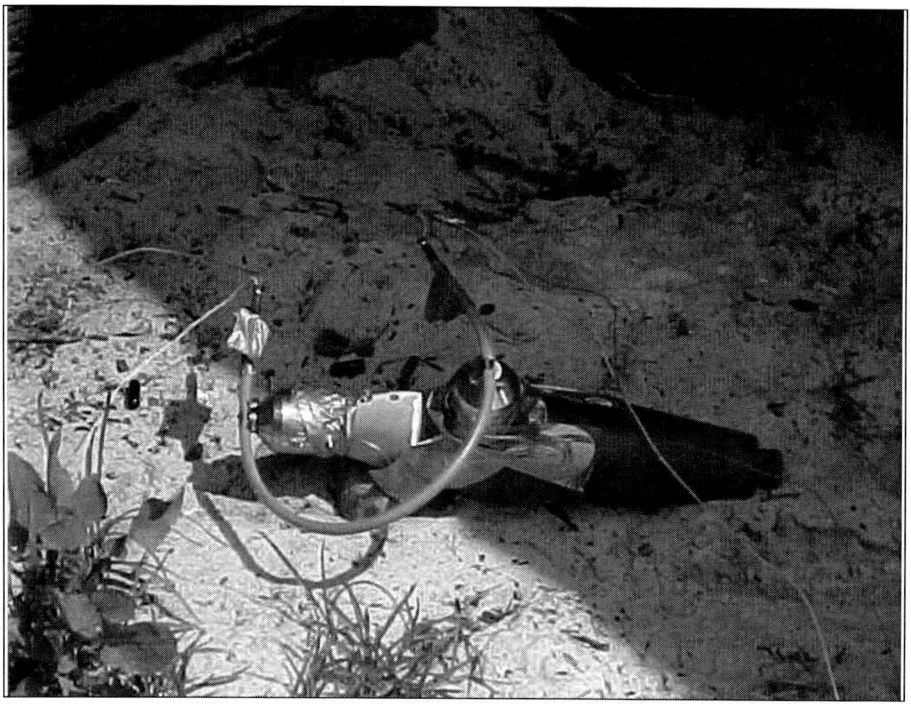

Figure 67. Bomb technicians attached to 48-gram shaped charges to disrupt the 81mm mortar round. The shaped charges are connected by 200-grain detonation cord and an electric blasting cap is attached to the detonation cord.

Since this jet must form, a formula exists for a "stand-off" distance for the charge. This allows the jet to reach its optimal shape and speed prior to impacting the matrix it will cut or penetrate. The mouth of the shaped charge provides a conical diameter with the typical stand-off distance about two to six cone "diameters." Some demolition experts prefer a 42° angle of the cone or opening of the mouth of the shaped charge. The depth of the liner material is usually about 3% of the thickness of the cone diameter if soft copper is used.

Another type of shaped charge is one that produces a projectile that has a very high speed. This IED is built with a six- to nine-inch steel pipe with one end welded closed with a hole drilled and a blasting cap inserted. Explosives are inserted in the pipe and an inwardly concave metal plate, usually copper, is placed over the open end of the pipe. The shock wave and hot gases propel the plate, sometimes called a platter, at very high speeds.

Figure 68. The mortar round following disruption by shaped charges. The round was not live.

Steel and copper can be used for the plate or platter, which forms a metal "slug" which is the explosively formed projectile. Speeds of 6,500 fps are common and the slug can penetrate up to five-inch-thick armor plate. The effective range is short, only about 15 meters for the optimal penetration ability, but the slug can penetrate some armor at 100 meters. With ranges less than four meters to the target, the slug does not form properly to allow optimal penetration. The IED is placed on the left side of the roadway to target the driver and may be at curb or guardrail height. Some are encased in concrete to conceal the IED. In many instances, a passive infrared device is the initiator when electronic countermeasures are used. These devices have been used effectively against US forces in Iraq and Afghanistan, using both passive infrared sensors and manual radio frequency devices to trigger the IED.

Chapter 7

INCENDIARIES AND OTHER EXPLOSIVES

Incendiary devices are called improvised incendiary devices or IIDs. These devices use fire and heat to destroy the target. They are easily constructed, require no training to construct, and can be deployed by hand, mail, or launched. These devices may be difficult to track as the ensuing fire may destroy most of the physical evidence. In some instances, a case of arson via an IID may not be detected as such. The open source terrorism material rates fire as a more substantial hazard with more destructive potential than explosives.

The construction of IIDs varies, but most will use a flammable liquid with a burning fuse delay. In many instances, the fuse will be a textile wick, which is ignited with matches, or a cigarette lighter and hand-delivered. In some devices, a frangible container is used to disperse the flammable liquid. This is the classic Molotov cocktail with a glass bottle filled with gasoline and a textile wick. However, IIDs can be very sophisticated using electronic timers, silicon rectifiers with battery power source to ignite an electric match or a device, which mixes two hypergolic substances when agitated (see Figure 69).

IIDs like IEDs will have a container, filler, and firing train. The container may range from frangible containers of glass, plastic jugs, or even plastic pipes. Instructions to manufacture an IID are available on the Internet and alternative literature. Fillers may range from flammable liquids to homemade napalm. Napalm is traditionally a thickened flammable liquid. When gasoline is used, styrofoam, rubber bands, or soap shavings can be used as a thickener. This allows the fire to "stick" to surfaces and burn longer, enhancing the destructiveness of the de-

Figure 69. Bomb technician Corporal Chris Vida throws a lit Molotov cocktail containing a combination of gasoline and rubber bands.

vice. Used motor oil is filler, which is thickened with organic compounds already and fine metal chips. Others include Thermite and Thermate as previously discussed.

More exotic IIDs may use hypergolic agents such as a mixture of HTH and brake fluid, potassium permanganate and glycerin, or a sugar-potassium chlorate mixture ignited by sulfuric acid. Many IIDs use a burning fuse delay, but some use other types of burning delays such as a cigarette inserted in a pack of matches or candle inserted in a pan of a combustible liquid, or incense sticks used in the same manner. Electronic timers can be used with a battery power source to ignite the device using an electric match or filaments from a low-voltage light source.

The more common IID is a Molotov cocktail, which comes in a variety of configurations. On version uses a hypergolic method to ignite the device. The standard gasoline is placed in a glass container along with sulfuric acid. Care must be used with this type device as some gasoline contains sufficient water such when sulfuric acid is added, the heat generated causes the container to fail igniting the device prema-

Figure 70. A reconstruction of a hypergolic device found by bomb technicians. The jar contained gasoline mixed with sulfuric acid and the outer newspaper was soaked in a sugar and potassium chlorate mixture. The device self-ignited on impact when the sulfuric acid mixed with the potassium chlorate sugar mixture in the newspaper. This device made an impressive thermal event.

turely. The container is wrapped in newspaper saturated with a sugar-potassium chlorate mixture. When thrown, the device shatters mixing the gasoline-sulfuric acid with the sugar-potassium chlorate material causing a hypergolic reaction and produces a very hot fire estimated in excess of 1,000° F. Another version reported in open source materials is a "booby trapped" Molotov cocktail. This device uses a glass bottle containing gasoline and sulfuric with a textile wick containing a zinc metal washer. When the device is moved the metal washer drops into the acid generating heat and pressure. In some instances, this may be sufficient to rupture and ignite the container (see Figure 70).

High Temperature Accelerants

High temperature accelerants (HTA) were first noted in several fires in Washington State during the 1990s. These fires are suspected of using a mixture of metals and oxidizers similar to solid rocket propellant. Fires of this nature are also suspected in Florida, California, and Canada.

The problem with these fires is the rapid evolution of heat even with little fire load. These fires also leave little forensic evidence. Water does not suppress the fire in the typical fashion as the burning metals may be accelerated by the high temperatures stripping off hydrogen atoms from the water.

Another concern is that these mixtures if contained could generate a low-explosive-type explosion. Further, the very high temperatures of these fires may cause early structural failure as temperatures in excess of 5,000° F have been reported. Firefighters encountering these fires have reported early flashover, rapid temperature rise in very short periods, pressurized smoke of a gray color pushing from the structures, bright flames, along with pyrotechnic-type particles. Two firefighter deaths are suspected as a result of HTA fires. Tests conducted by the Seattle, Washington Fire Department in concert with the BATFE, US Navy, Lawrence Livermore National Laboratories, and Sandi National Laboratories had difficulty identifying the composition of the materials used.

Some observations were steel and iron structural components melted and vaporized while concrete spalled and showed greenish-blue discoloration. Tests by the Seattle Fire Department demonstrated one-hour fire rated walls failed within five minutes of the fire start when HTA is involved. Tests have shown this is not a thermite mixture but more akin to a solid rocket propellant. A full-scale HTA test using roughly 450 pounds of a similar compound on an abandoned building 250 feet by 120 feet with little fire load, involved the building fully within three minutes of ignition. Speculation is the HTA material is a combination of ammonium perchlorate, ammonium nitrate, aluminum powder, plus diesel fuel.

Found Explosives

A common issue facing bomb squads is found explosives. Each year bomb technicians are injured handling and disposing of found explosives including fireworks. This may vary from stolen explosives to old deteriorating explosives found from agricultural or mining uses. When dynamite was in common use, bomb squads faced a common issue of old dynamite. Although rare now, old nitroglycerin, or straight dynamite, was a common issue in the 1970s (see Figures 71 & 72).

Nitroglycerin dynamite becomes unstable with age as the dynamite "sweats" nitroglycerin. Nitroglycerin is sensitive to friction, shock, heat, and flame. Movement may cause detonation. On rare occasions,

Figure 71. Several hundred pounds of old ditching dynamite recovered from a farm. The shed with the dynamite was burned. The dynamite did not explode.

Figure 72. This ditching dynamite was sufficiently degraded that bomb technicians burned it and the barn containing it. The dynamite did not explode.

Figure 73. This old dynamite in an emulsion form was safe to move and was burned outside the barn where found. The dynamite did not explode.

nitroglycerin may be found. This is usually a homemade version. Nitroglycerin is usually a clear to yellowish oily liquid. As it oxidizes and becomes unstable, it becomes a reddish color.

Movement of nitroglycerin dynamite and or nitroglycerin itself is dangerous. Many bomb squads elect to burn nitroglycerin dynamite in place. Nitroglycerin liquid, which is old, will likely detonate if moved. Some individuals have elected to use a desensitizing agent such as acetone to mix with the liquid nitroglycerin to allow movement. This risky procedure has on more than one occasion resulted in an explosion. In one recent case, a bomb squad faced disposing of a 125-year-old bottle of nitroglycerin found in a storage shed. The bomb squad elected to burn the nitroglycerin in place and burned the shed. In another event, a bomb squad in northwest Florida dealt with almost a ton of deteriorating dynamite when the farm owner died and relatives discovered the dynamite. The dynamite was burned in place, which included burning the barn where the dynamite was located. No explosion occurred. The extra dynamites using ammonium nitrate and other nitrate compounds are not as sensitive as nitroglycerin-

Figure 74. This dynamite was mixed with loose fuse-ype blasting caps. Bomb technicians burned this dynamite and shed containing it. The dynamite did not explode, but the blasting caps did explode.

based products. These can usually be moved and burned without undue danger (see Figures 73 & 74).

Old blasting caps present a tremendous danger. These caps, especially copper metal blasting caps, form very unstable peroxide compounds. Friction or movement easily ignites these compounds. These peroxide compounds are also common in containers of old ethyl ether. The simple turning of the lid may be sufficient to detonate the ether. Picric acid forms a similar very friction-sensitive compound when exposed to metal, making picric acid in metal containers or glass containers with metal lids very unstable. Japanese pottery grenades along with some WWII shells use picric acid as filler and thus are very sensitive to movement or friction (see Figures 75 & 76).

Another issue facing the bomb technician is the disposal of illegal fireworks. These along with home-based fireworks manufacturing laboratories are very dangerous. Flash powder is very static electricity sensitive and movement within such a laboratory may result in an explosion. Some fireworks use illegal mixtures of mercury fulminate, which is a very unstable primary explosive.

Figure 75. Bomb technicians were called when these old blasting caps were found. These blasting caps were destroyed by countercharging in place with a small amount of C4.

Figure 76. These fuse type-blasting caps were destroyed by burning the shed that contained them. They did explode.

Figure 77. Three men were killed here because they struck a drill tube filled with 300 pounds of dynamite with a sledgehammer in an effort to loosen it, which then exploded and detonated a truck containing 3,000 pounds of dynamite. (From Thomas Brodie, *Bombs and Bombings*, 3rd ed. Courtesy of Charles C Thomas, Publisher, Ltd.)

Other common modern explosives found include modern dynamite, detonation cord, boosters, shaped charges such as oil well perforators, agricultural explosives, and other proprietary explosives such as squibs. Although most modern explosives are stable, even some stable compounds such as TNT become unstable when in contact with metal. Another issue is the toxicity of explosives such as TNT. Even modern explosives can injure, as a bomb technician was recently injured handling oil well perforators.

Each year, individuals are injured dealing with explosives that have been exposed to heat. Heat may sensitize explosives making them sensitive to shock or friction. In one circumstance, the impact of hammer on a drill, which had inadvertently drilled into a mining column of dynamite, caused an explosion and death of several blasters (see Figure 77).

Military Ordinance

A common issue is found military ordnance. Many individuals from WWI, WWII, Korean War, Vietnam War, Desert Storm, and the current conflicts have brought a variety of ordnance home. Many of the WWII and Korean War veterans are dying, which means family members sorting through personal property are finding explosives, hand grenades, shells, blasting caps, Claymore mines, and other assorted military items.

Items smuggled off military facilities, stolen, lost, or recovered from impact areas are common. Even shells from the Civil War are still live in many cases and dangerous. Each year Civil War enthusiasts collect and attempt to demilitarize Civil War ordnance. A search of the Internet found several news stories where individuals were injured and killed by such efforts. By law, the US government owns the military ordnance regardless of the manner the individual acquired it. The US Army reports that more than 16,000 inactive ranges with impact areas containing unexploded ordinance exist. The issue is so serious the Department of Defense has established a website, http://www.denix.osd.mil, for the public to use for safety regarding unexploded ordinance (UXO). The website reports several fatalities from handling UXOs and finding UXOs in a variety of locations which included construction sites near military facilities, abandoned residences, scrap metal yards, museums, near Civil War battle sites, and along the coast. In several of the instances, children who did not recognize the danger of the ordnance were injured or killed when it detonated (see Figures 78-81).

In many instances, the bomb squad will identify the item as military ordnance and determine the item is likely live, notify the nearest military explosive ordnance disposal (EOD) unit, and follow their recommendations. In many cases, the military EOD unit will respond and assume control of the incident and dispose of the item. Use of the aforementioned website for identification purposes is helpful as the observer can compare the types of ordnance to what was seen.

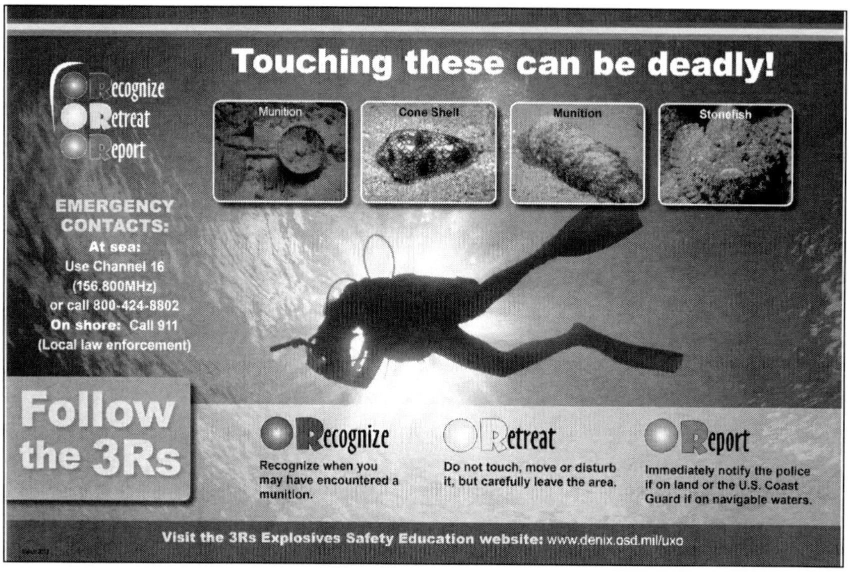

Figure 78. Follow the 3Rs of Recognize, Retreat, Report. Many UXOs can be found underwater or along shorelines.

Figure 79. Another UXO chart showing the 3Rs for unexploded ordinances found underwater or along shorelines.

Figure 80. Many UXOs can be found near military facilities.

Figure 81. Live modified hand grenades. These items were found while answering an unrelated call. (Photograph by bomb technician Corporal John Skipper.)

When feasible the person or officer observing should be asked if it looks like any of the following:

- Grenade
- Mortar round
- Artillery projectile
- Aerial bomb
- Small arms ammunition

If the observer cannot identify the ordnance, perhaps they can describe the body of the item:

- Cylindrical
- Cone shaped
- Spherical or ball shaped
- Pear shaped
- Rod shaped

Other helpful information includes color, any markings especially colored bands, any fins, safety pins or spoon, or a spike on the nose of the ordnance.

Some Civil War era ordnance is still live and dangerous. Live shells have been found on battlefields, in collections, at antique shops, and in museums. Civil War shells and munitions contain black powder. The black powder in the devices may be live. Many of these devices will still be live even though they may have been buried or submerged since their use. Some may have live fuzes, which contain very unstable mercury fulminate. Some have percussion caps in the nose such that impacting the nose could detonate the shell. Each year several amateur collectors are injured while attempting to demilitarize these shells. The typical method is to drill through the casing and remove the black powder. The heat and friction of the drilling ignites the black powder and causes an explosion. The US government is still the owner of these shells and munitions. Bomb squads near Civil War battlefields or naval battles deal with these shells and munitions. Some of these items are found in fishing nets or on beaches near naval battle sites. Some Civil War ordnance experts cite a "dud" rate of 40% to 70% depending upon the type projectile.

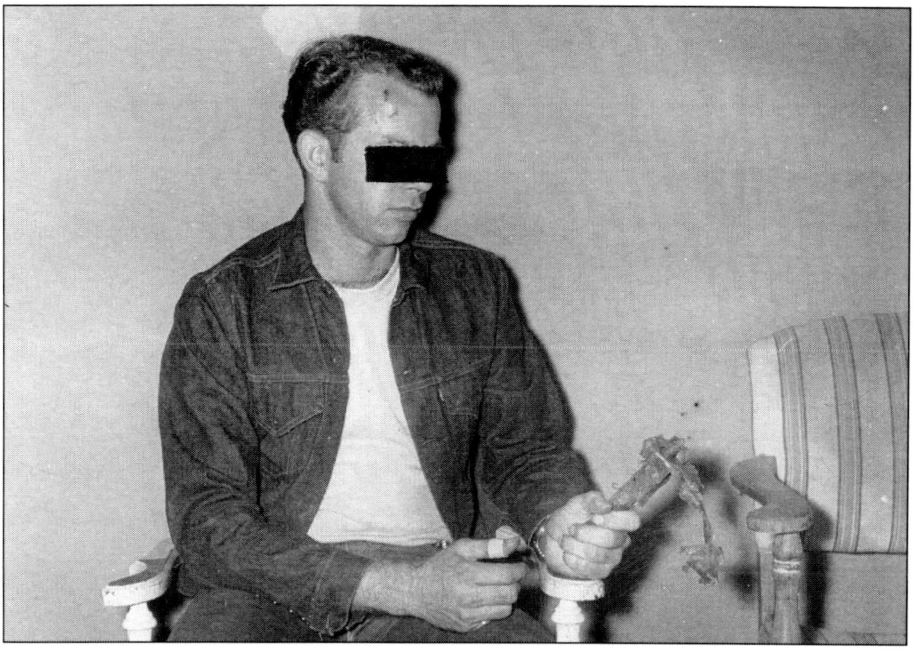

Figure 82. A Dade County citizen found this white phosphorous rifle grenade in a World War II ammo dump, broke it open, burned phosphorous, and broke open the burster tube, which burst on him, cutting his thumb and forehead. (From Thomas Brodie, *Bombs and Bombings*, 3rd ed. Courtesy of Charles C Thomas, Publisher, Ltd.)

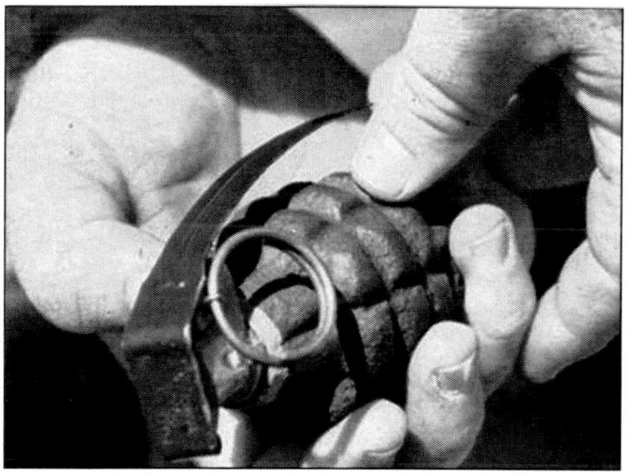

Figure 83. This live hand grenade from WWII was recovered from a reenactor who had a collection of ordnance.

The projectiles can in a variety of configurations and a large number of variations exist because of the numerous manufacturing sites. Bolts are solid shot, which may appear in the traditional cannonball configuration, or the conical shape of a cannon projectile. Shells contain explosives and are usually hollow and but might be spherical or polygonal in shape. These were time fused and impact fuzed. Case rounds were usually spherical and contained additional fragmentation. These were designed for airbursts against troops in the open. Canister or grapeshot has many metal balls in a container within a container and was used against troops with direct fire.

States in which projectiles have been located (partial list):

Alabama
Georgia
Louisiana
Mississippi
Missouri
New York (Revolutionary War projectiles)
North Carolina
South Carolina
Tennessee
Texas
Virgina
West Virginia

Battlefield where sites projectiles have been located (partial list):

Atlanta, Georgia
Augusta, Georgia
City Point, Virginia
Chattanooga, Tennessee
Corinth, Mississippi
Coosawhatchie, South Carolina
Dalton, Georgia
Fort Davidson, Missouri
Fort Pillow, Tennessee
Fredericksburg, Virginia
Harpers Ferry, West Virginia
High Bridge, Virginia
Honeyhill, South Carolina

Hurricane Creek, Mississippi
Kennesaw, Georgia
Kolb's Farm, Georgia
Macon, Georgia
Manassas, Virginia
Milledgeville, Georgia
Mobile, Alabama
Nashville, Tennessee
Oconee River, Georgia
Pensacola, Florida
Petersburg, Virginia
Port Hudson, Louisiana
Resaca, Georgia
Richmond, Virginia
Savannah, Georgia
Shiloh, Tennessee
Spanish Fort, Alabama
Spotsylvania, Virginia
Stones River, Tennessee
Sulphur Trestle, Alabama
Shiloh, Tennessee
Vicksburg, Mississippi
Welch's Cemetery, Louisiana

Armory, storage, or manufacturing sites (partial list):

Atlanta, Georgia
Augusta, Georgia
Marshall Powder Mill, Texas
New Orleans, Louisiana
Richmond, Virginia
Robeline, Louisiana
Selma, Alabama
West Point, New York

The author has had to deal with hand grenades, aerial bombs, TNT and C4 explosives, Claymore mines, white phosphorus smoke grenades, CN and CS grenades, mortar rounds, 40mm grenades, missiles from shoulder fired and aircraft launched apparatus, 37mm, 105mm

Figure 84. This item was dredged from the Mobile Bay in Alabama and was identified as a shell from the Civil War. (Photograph by Sergeant Gary Coleman.)

and 75mm shells, and land mines from WWI, WWII, Korea, and Vietnam. The current conflict in southwest Asia has produced similar findings including many foreign hand grenades and plastic explosives. The author even found a live hand grenade on a desk of secretary in a healthcare facility. Her solider boyfriend provided the grenade. The regional airport has requested assistance on multiple occasions with military service members attempting to board aircraft with military explosives and military ordnance secreted in their luggage. Some areas have issues with shells and military munitions washing ashore or in nets of fishing vessels in the areas of naval battles or ocean disposal sites (see Figures 82–84).

Another problem is civilians purchasing demilitarized hand grenades, filling the bodies with smokeless gunpowder and using a smoke grenade fuse as the igniter. In many instances, the grenade is functional, albeit not as potent as when high explosives are used, and hazardous. These can be found in some clandestine drug laboratories. The problem is, the smoke grenade fuse ignites instantaneously when the safety spoon is released. The resulting explosion occurs while the grenade is still held. Some have been used successfully as booby traps using the instantaneous explosion to an advantage while others have had injuries from the exploding device. One bomber used the smoke grenade to ignite a short piece of cannon fuse attached to a homemade blasting cap inserted into the hand grenade body. One crafty bomber used dynamite within the grenade body but had a low order explosion when the old dynamite failed to ignite from the explosion of the homemade blasting cap.

Bottle Bombs

An unusual hybrid device, a favorite of juveniles, is a bottle bomb. These devices operate from overpressure from a chemical reaction. Some have reported these devices on rare occasions will produce a thermal event. Common fillers include isopropyl alcohol mixed with HTH, aluminum foil mixed with hydrochloric, or muriatic acid, and dry ice placed in a bottle. The chemical reaction of the acid and metal or alcohol and HTH is to produce a hot gas. In many instances, a corrosive liquid will also be present. When the container fails, the hot gas escapes propelling the hot liquid with it. If glass is used, glass fragments are produced and propelled. The devices using dry ice are a simple overpressure device as the dry ice sublimates to carbon dioxide gas, the container will usually fail. Multiple variations exist and the Internet has many formulas and construction methods available.

Juveniles use these devices as an experiment and as a destructive device. These devices appear nonthreatening, but they have sufficient power to destroy a mailbox and if exploded on the ground, leave a small crater. Eight-ounce glass bottles using aluminum foil and hydrochloric acid have propelled glass fragments 100 feet. The failing point on many containers will be the plastic cap. This usually fails above 100 psig. These devices will destroy mailboxes.

The filler variation and container variation make these devices unpredictable as to when they will explode. In many instances, juveniles learn in chemistry class how to evolve hydrogen gas using mossy zinc and hydrochloric acid. Any capped bottle with a muddy appearing liquid or metal strips or fragments should be considered a bottle bomb. This can be confused with a "shake and bake" methamphetamine reaction, which also produces overpressure in capped bottles from a similar reaction. In both cases, the bottle is likely to fail if the cap is intact.

Bomb squads take several approaches to rendering the device safe. Some use pellet guns to rupture the container while others use a dis-

Figure 85. This bottle bomb was thrown at a pedestrian by a passing car. The pedestrian was fortunate and not injured in the ensuing explosion. Note the unconsumed aluminum foil in the photograph.

rupter. Some go to the length to use a thermal imager and remote infrared thermometer to see if the device is still evolving heat. If the device is not evolving gas, the bomb squad will use the bomb squad robot with a sharp implement to puncture the plastic bottle and allow the gases to escape. This is a substantial amount of effort for device that will likely not produce any physical evidence. The corrosive liquids used in many of the bottle bombs will obviate the chance of recovering fingerprints from the container in many cases. In some locales this device is treated as a destructive device, others treat the devices as a bomb, while others consider the device an incendiary. In a few locales, the law does not address this device as illegal (see Figure 85).

With detonated bottle bombs, one will likely find dead vegetation from the corrosive liquid. Collecting the device can be problematic, as the device likely will be covered with a corrosive liquid. Some agencies collect the remnants after the items have dried. However, even in a dry condition, the items will likely be corrosive.

If what is suspected to be a bottle bomb, the item should not be moved but should be isolated, then evacuate 100 feet in the open, and stay uphill and upwind. Remember, these items should be not approached or moved. A police officer in the southeastern US had permanent hearing loss caused by the detonation of one of these devices. The patrol supervisor was checking mailboxes in a residential area for

Figure 86. This mailbox, advertised as indestructible, was destroyed by a 10-ounce glass bottle bomb. Note the aluminum foil scattered around the mailbox. Glass fragments were projected more than 100 feet.

bottle bombs when the mailbox she drove up to exploded. This was one of several mailboxes targeted by young adult offenders. The mailbox contained a 10-ounce bottle bomb, which exploded. The officer lost hearing in the ear facing the mailbox (see Figure 86).

Chapter 8

OTHER WEAPONS OF MASS DESTRUCTION

Preparation and Response to Terrorism

Terrorism is a fact that responders must adapt to and consider within response procedures. The more important part of the response to WMD events or terrorist attacks is realizing such can occur, adjusting plans and procedures to meet such events, and early detection. These events may require PPE and modified tactics. These events will likely cause mass casualties. Weapons of mass destruction can be described by the acronym CBRNE. This includes chemical agents, biological agents, radiological agents, nuclear agents, and explosives, which includes incendiary devices. One must also consider firearm attacks, hostage taking, and cyber attacks as part of the threat spectrum.

Perhaps the more fundamental approach to this new environment is to assure that the responder is safe and able to respond to an event. A review of the security of the facility that houses the responders is essential. Vigilance and the ability to recognize an attack has occurred is another essential skill. The responder must analyze what targets are within his or her jurisdiction. The change by terrorist from well-guarded hard targets to soft targets such as hotels, schools, hospitals, rail transportation systems, or densely occupied events has been deliberate. The responder must also consider such infrastructure, which could be attacked that would impair the responders' ability to complete their mission along with causing disorder. Power failure over a large area is one method to cause disorder, as this would be disruption of communications and cause traffic issues. If the terrorist wished to amplify an attack, they might attack infrastructure to cause a loss of power and commence attacks against multiple targets such as the attacks in Mumbai. Most agencies in the US are not prepared for such an eventuality. The North Hollywood shootout, the Dorner attack, and the Norco bank robbery demonstrate that a few "shooters" can exhaust law enforcement resources over a large area. Even low intensity attacks such as those carried out by UNABOMBER can create terror out of proportion to the threat and evade law enforcement while requiring the commitment of substantial resources to the counter the attack.

One failing in many events is prompt notification of the chain of command and recall of off-duty personnel. Even the requesting of outside resources may be impaired, as was seen in Katrina. Further, responders usually do not have a coherent evacuation plan for tactical emergencies and strategic evacuations of the agency. Although facility security has improved, the advent of VBIEDs and the use of multiple attackers, as seen in the US embassy bombers, attacks in Mumbai, and Paris can defeat defense systems even when staffed with several armed personnel aware of the threat.

If chemical agents are deployed, responders may have to don PPE. In many instances, the practice and presence of PPE on the responder's vehicles is an issue. The first responders are likely to become casualties. This will drain resources in rescue attempts of downed responders and demoralize public safety personnel. Further, responders may have to self or buddy aid with nerve agent antidote kits. Chemical, biological, and radiological agents require the use of respiratory protection. This is likely to be in the form of an air-purifying respirator. These agents are usually dispersed as dusts, mists, vapors, or aerosols. Few agencies have respiratory protection programs and issue APRs to field personnel.

In the typical setting, responders will likely not have access to Level C attire. This is the lowest rated PPE. This level provides skin protection and purifies air through filters. The shortcoming is the agent must be known such that the filter can be confirmed as rated against the threat and the atmospheric oxygen levels must be normal. The attire is not effective against vapors. Level B is similar but uses a self-contained breathing apparatus. This provides an enhanced respiratory protection, but the attire is usually not effective against vapors. The ultimate protection is Level A that uses an encapsulated suit, which provides protection against vapors.

Use of PPE decreases the ability to see, hear, move, and increases heat load. In many circumstances, the time in PPE may be limited to less than one hour. The need to change air bottles in self-contained breathing apparatus is one limiting factor. Cooling vests may decrease the heat load. These factors can cause officers to be impaired in tactical operations as this additional weight and loss of situational awareness caused by the PPE. This will reduce their effectiveness.

Officers will need to wear body armor in every environment. The body armor must be matched to the threat. Level II body armor defeats many handgun rounds whereas Level III defeats some carbine rounds, while higher level with ceramic and metal inserts are used to defeat rifle rounds. The issue is the higher the threat level protection, the more rigid and heavy the body armor becomes. Bomb suits may weigh in excess of 90 pounds. Body armor will impair mobility substantially.

The index of suspicion when responding to calls, which might be a terrorist-related event or be active aggressive behavior from normal criminals, should be raised to consider such events as high risk. Items which should gain the responder's attention as a potential high-risk event, terrorist event, or beginning of a parallel attack of multiple sites, include active aggressive behavior with shooting, reports of a hazardous materials release, explosions, and reports of persons down under unusual circumstances. If the reports of persons down includes persons dead or seizing, a nerve agent attack should be considered. Responses to suspicious powders from mail or dispersed at a location, especially a government building should prompt the responder to consider a biological, chemical, or radiological attack.

As responders near the scene, they should have obtained the available from dispatch and approach cautiously. Some agencies shutdown emergency equipment several blocks away to allow a quiet and observing approach to scene and never presume the information from dispatch is correct. If the scene appears to be unstable, stage outside the danger area. Remember if you can see the scene, you are within gunfire range and bomb fragment range.

If the reports of persons down are the type call, consider approaching from upwind, avoiding mists, fumes, smoke, and vapors. Be observant for out of place persons, vehicles, packages, or other suspicious items. Persons trying to hide themselves may be a threat. These may be secondary devices, suicide bombers, or persons present to attack responders. Responders should use caution in the presence of crowds or groups of disorderly persons or when the scene has no one present.

When siting a command post for an event, pick the best location and discard it as it might be targeted by attackers with prepositioned IEDs or an ambush with firearms. Remember to set up the command post upwind and uphill from the event when feasible. Establish egress, exits, hot zones, warm zones, and cold zone if the event involves explosions, potential chemical, biological, or radiological items. Establish incident command and use the incident management system to control the event.

If numerous persons are injured, once an area is designated, have those who can walk out of the scene do so. If they are potentially contaminated, they will have to be assessed by EMS personnel wearing at least Level C PPE. Only enter the scene to render lifesaving aid and then use expedient methods to remove those persons injured from the scene. Wear PPE if the potential for contamination exists.

EMS will likely establish a triage area to sort those patients by priority. Red tags indicate an immediate need for emergency medical care off site while yellow tags indicate the transport can be delayed. Green tags are minor injuries. Black tags are for those already dead or beyond salvage. Leave the dead in place and do not expend resources or enter a danger zone to move bodies.

Chemical Agents

Chemical agents have been in use for hundreds of years. They can be manufactured in a noncommercial setting and are lethal. Early chemical agents used naturally available components and were crude. Poisoned arrows, smoke, burning arsenic or sulfur, and the like are early examples. World War I saw the advent of more sophisticated agents starting with toxic industrial chemicals such as chlorine and phosgene. Vesicants followed shortly. Open source records show about 90,000 persons were killed in WWI and more than one million injured by chemical agents. Chemical agents are useful against crowds or in areas of dense population (see Figure 87).

Modern uses of chemical agents include the alleged use by Iraq against the Kurds and Syria's alleged use against rebel forces. Iraqi forces are suspected of killing several thousand civilians using a mix of chemical agents. The agents suspected include nerve agents, vesicant

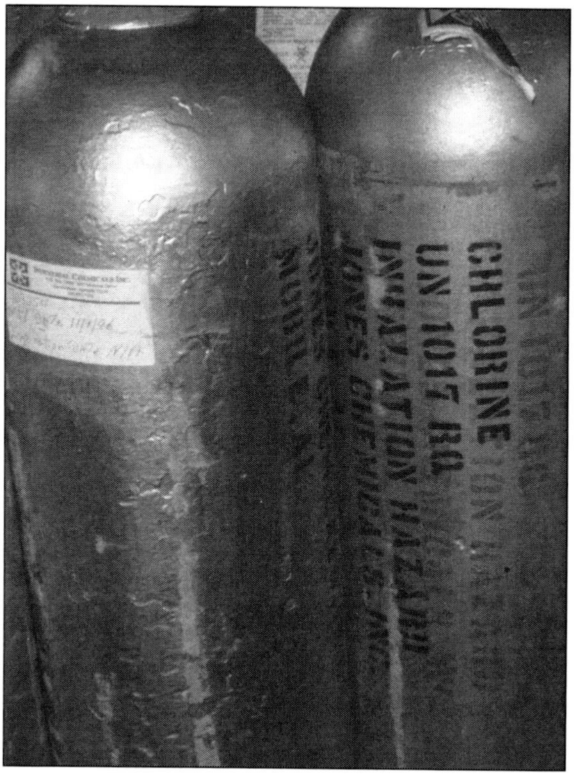

Figure 87. A 100-pound cylinder of chlorine.

agents, and hydrogen cyanide. Sarin nerve agent was the agent used in the 1995 Tokyo subway attacks and the attack in Matsumoto by Aum Shinrikyu, a radical religious group. One issue seen with this attack was the off gassing of vapors from the contaminated victims and their clothing exposing responders and medical personnel. Many responders and medical personnel received nerve agent exposure through this method. In fact, during the 1994 sarin attack in Matsumoto, responders and medical personnel were not certain what killed the eight persons in the apartment complex attack. Further, they were not certain what caused the effects of the hundreds sickened by exposure to sarin. One issue was the lack of training and information regarding the threat posed by Aum Shinrikyu with these agents. This was a dress rehearsal for the 1995 attacks on the Tokyo subway. Unconfirmed reports state recently that the suspected use of a binary artillery shell detonated near a US convoy in Iraq produced symptoms of nerve agent exposure to some members of the convoy.

Chemical agents are moderately difficult to manufacture, but their use causes terror out of proportion to their direct effects. If the chemical agents cannot be manufactured, toxic industrial chemicals could be substituted. This might be chlorine stolen from water treatment plants, anhydrous ammonia acquired from agricultural operations, or illegally obtained organophosphate insecticides used in lieu of nerve agents. Although no longer manufactured, a very toxic agent, which kills through organophosphate poisoning which binds acetylcholinesterase, is Temik®. It contains the pesticide aldicarb. This is a rapid-acting pesticide, which is toxic to mammals, and low doses can be fatal. This could be used as a substitute nerve agent.

Issues with chemical agents are that these agents are usually dispersed as an aerosol. This means that wind direction is critical in determining what areas will be affected. Some of the agents such as cyanide are quickly dispersed while others such as VX nerve agent and vesicant agents are persistent. Chemical agents such as chlorine, which is heavier than air, tend to accumulate in low-lying areas. This makes siting the command post and staging areas upwind important. If any indoor attack occurs, ventilation of the structure will be important, but one must remember that releasing some nerve agents into the external atmosphere may affect persons downwind.

Decontamination of persons exposed to chemical weapons will be an issue. Do not allow contaminated persons to approach responders unless the responders have respiratory and skin protection of at least level C donned. Doffing contaminated clothing will remove a substantial portion of contaminants. A rinse with water, particularly if coupled with use of soap on the hair, will remove essentially all the contaminants.

Nerve agents were developed from pesticides and were developed by the German Army. In fact, the nerve agent's official designation such as GA, GB, GD, and GF stand for German and the letter is the order they were developed. GA is known as Tabun, GB is Sarin, GD is Soman, and GF is Cyclosarin. These agents have a volatility similar to water. VX, developed in the United Kingdom in the 1950s from the insecticide parathion, is more akin to oil and volatilizes slowly. VX is one of the more poisonous manmade substances today with as little as 10 mg being lethal with skin contact. This is a drop, which is barely visible.

Chemical agents use a variety of methods to injure and kill. The more potent nerve agents attack the nervous system causing respiratory arrest and death. Nerve agents are a terror weapon as they kill rapidly, can injure hundreds, but the negatives are they require technical skill to manufacture and their dispersal can be technically difficult. The typical exposure to nerve agents will be through inhalation. This produces shortness of breath, chest tightness, eye pain, and dimness of vision. If concentrations are high, the person may soon collapse and suffer seizures followed by death from respiratory arrest. Another consideration is persistence. The US Army defines chemical agents, which are active more than 24 hours as persistent. Most nerve agents such as sarin have properties similar to water and tend to be nonpersistent. Nerve agents in the "V" series and vesicant or blister agents tend to have an oily consistency and are persistent. These agents are useful for area denial as they persist more than 24 hours with VX lasting from three to eight days depending on temperature and weather conditions. Sarin typically lasts from 30 minutes to eight hours depending upon environmental conditions. Nerve "gas" is a misnomer as the agents are aerosols or liquid droplets. Environmental conditions affect their dispersal, persistence, and effectiveness. Some are hydrolyzed by rain or moisture; UV light in sunlight destroys some, while others are diluted by winds. Dispersal will depend upon temperature and winds. However, with persistent agents protected from sun and water, they can last a longer time and some vesicant agents from WWI are still dangerous.

Nerve agents are preferred as they are lethal in small quantities and kill quickly. Nerve agents are derived from organophosphate pesticides, which inhibit cholinesterase production and use allowing the "vegetative functions" of the nervous system to stop respiration. Organophosphate pesticides could be used as an improvised chemical agent. Most nerve agents will be dispersed as an aerosol or droplets. This could be a handheld sprayer, industrial foggers, aircraft, or by explosion. Therefore, any explosion should be considered as a method to disperse a chemical or radiological agent.

Some of the more common nerve agents are:

- GA or Tabun
- GB or Sarin (named after its inventors **S**chrader, **A**mbros, **R**udiger, Van der l**in**d)

- GD or Soman
- GF or Cyclosarin
- GV (semi persistent intermediate toxicity)
- VX and V series (very poisonous, persistent oily agent)

Table 4. Toxicity Data

Agent	*Effect*	*Ct50 (mg-min/m3)*
GA - Tabun	miosis	2 - 3
	death	200 - 400
GB - Sarin	miosis	3
	death	100 - 200
GD - Soman	miosis	2 - 3
	death	50 - 70
VX - Venom	death	10 - 50
G" Agents		

Lethal Dose 50% of those Exposed on Skin

GA - Tabun	1,000 mg (1 gram)
GB - Sarin	1,700 mg (derived from parathion)
GD - Soman	50 mg
GF - (no common name)	30 mg

The more common exposure method will be through respiration. Exposed skin can absorb the agent in liquid droplets. Indicators of nerve agent exposure in vapor form will be difficulty breathing (dyspnea), eye pain, pinpoint pupils (miosis), and running nose (rhinorrhea). A substantial exposure can cause loss of consciousness, seizure, and respiratory arrest. These signs and symptoms occur quickly upon vapor exposure. Skin exposure to droplets causes perspiration at the point of exposure, nausea and vomiting, muscle twitching (fasciculation) at the site of exposure, weakness followed by rhinorrhea, and lung secretions. Death may follow quickly. Droplets on the skin take longer to cause these signs and symptoms. Lethal amounts with skin contact vary from around 4,000 to 300 milligrams for G agents to roughly 10 mg for VX. Aerosol and vapor deployments also follow a similar toxicity pattern. Sarin is said to be more than 25 times more poisonous than hydrogen cyanide. These agents may be delivered in the finished form called a "unitary" deployment. Some delivery methods require the mixing of the two components to deliver the final nerve agent. This is called a "binary" delivery. The danger from a unitary delivery is the inadvertent release of the agent as the binary agents are safe until mixed. The issue with binary agents is thorough mixing must occur for the agent to have sufficient purity to function properly. The aforementioned alleged deployment in Iraq was said to be a binary delivery with an IED used to detonate an artillery shell. The explosion appears to have not sufficiently mixed the two precursors to produce sufficient agent to affect the convoy members.

*Nerve Agent Precursors**

- dimethyl methylphosphonate
- phosphorus trichloride
- sodium fluoride
- alcohol

Signs of Nerve Agent Deployment

Since nerve agents are likely to be used in crowds or areas of dense populations, signs of these agents use are:

- Numerous persons down from no apparent reasons
- Several persons down with seizures
- Large number of persons complaining of eye pain
- Large number of persons complaining of difficulty breathing
- People with pinpoint pupils and rhinorrhea
- These signs at a high value target or following an explosion

Remember, these individuals are likely contaminated and should not be contacted without chemical protective clothing and at least an air-purifying respirator. Many of the responders in the Tokyo subway attacks were injured by the off gassing of sarin nerve agent from the clothes of those contaminated in the attack. Some authorities cite the

---

*Several open sources claim that these items are purchasable via the Internet.

Figure 88. Older Mark One nerve agent antidote kit carried on web gear.

off gassing of clothing will occur for at least 30 minutes. The area likely contaminated should not be entered without appropriate respiratory protection and chemical suit. Nerve agents can be detected using commercial sensors, military M8 and M9 test paper for liquids, and M256 devices, pesticide ticket detectors, and other colorimetric tests. If a nerve agent is indicated, personnel should immediately don respiratory protection and chemical clothing. The North American Emergency Response Guide 153 should be followed.

Use time, distance, and shielding to protect oneself. Spend minimal time in the hazard area, stay upwind and away for the site, and shield oneself with respiratory protection and a chemical suit for skin protection. If public safety personnel are down or exposed, the auto-injectors supplying atropine and 2PAM chloride should be used if these individuals are symptomatic. These may be in the older Mark One kit or single DuoDote® injector. Some agencies issue these items to public safety personnel. EMS agencies will not carry sufficient stocks of atropine for this purpose. The typical number of injectors is three per responder for self or "buddy" use. These kits should only be used if the responder is symptomatic and exhibiting substantial signs of nerve agent exposure such as dimness of vision, eye pain, difficulty breath-

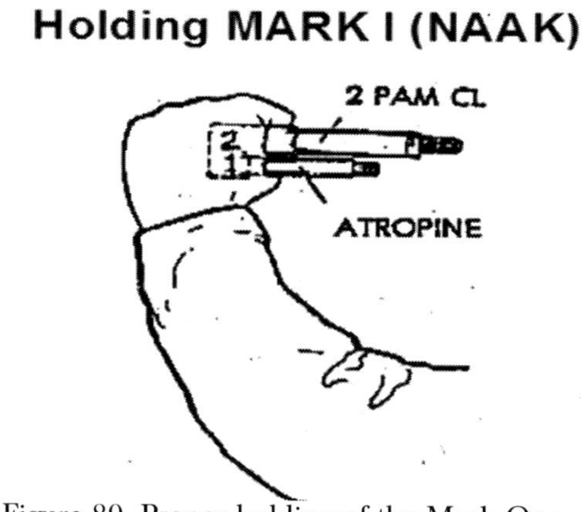

Figure 89. Proper holding of the Mark One.

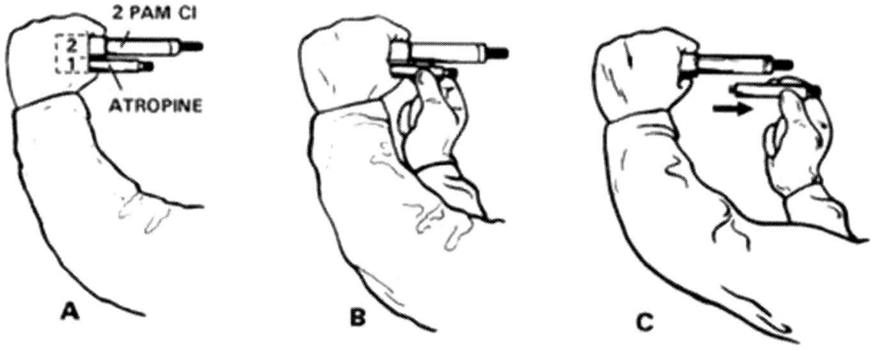

Figure 90. Proper removal of the Mark One.

ing, and rhinorrhea. Remember the removal of clothing will likely remove much of the contamination. The kits can be injected through a single layer of clothing. The older NAAK contains two auto injectors with one containing two mg of atropine while the other auto injector contains 600 mg of pralidoxime (2-PAM). The atropine reduces the effect of the nerve agent on the body's vegetative system by raising heart rate, dilating bronchial tubes, drying secretions, and alleviating intestinal spasm, while the 2-PAM breaks the bonds of the nerve agent. The DuoDote® injector delivers the same drugs sequentially through a single needle. The injectors should be used on the outer portion of the thigh about midway between the knee and hip. An alternate site is the upper buttocks, avoiding the middle of the buttocks where the sciatic

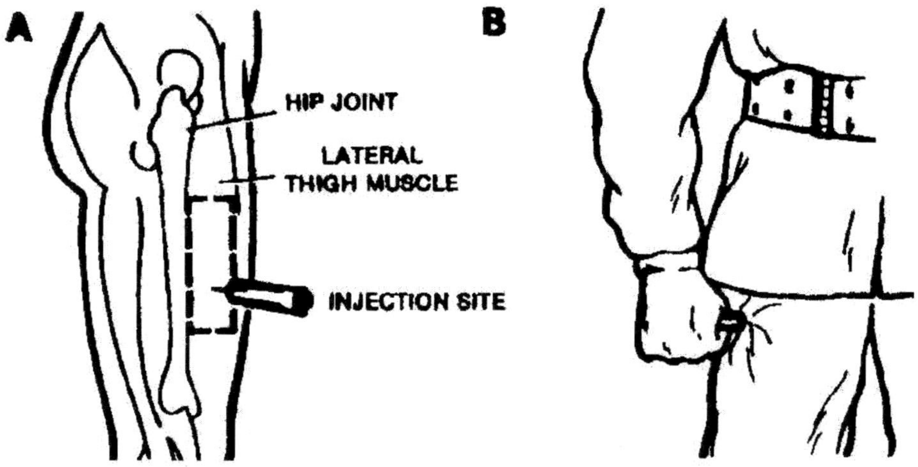

Figure 91. Lateral thigh injection sites and self-injection.

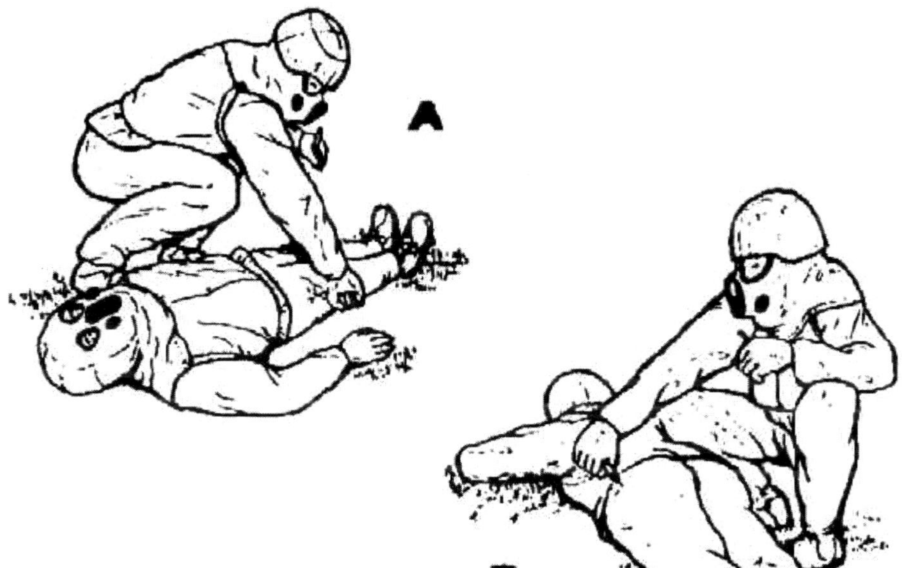

Figure 92. Buddy aid injection using the Mark One.

nerve passes. Once used, the injector will have a needle exposed, which must be treated as a sharp. Disposal in a sharps container or bending the needle to reduce the sharp hazard is recommended (see Figures 88–94).

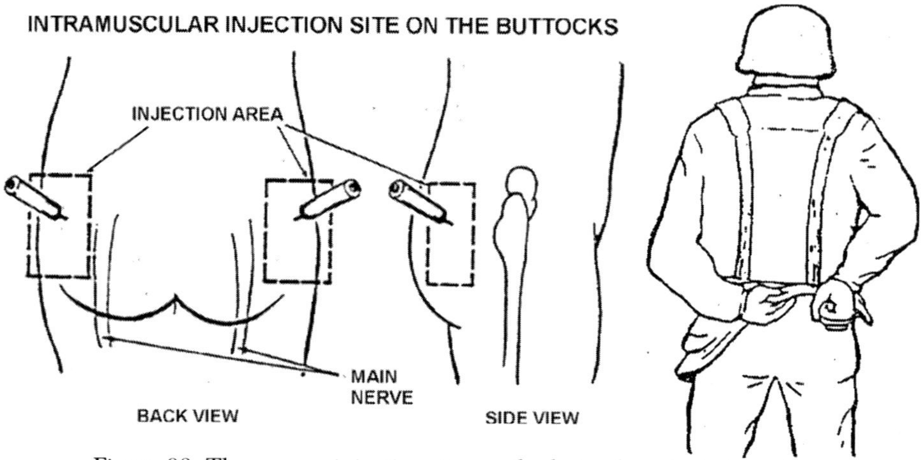

Figure 93. The proper injection site on the buttock and self-injection.

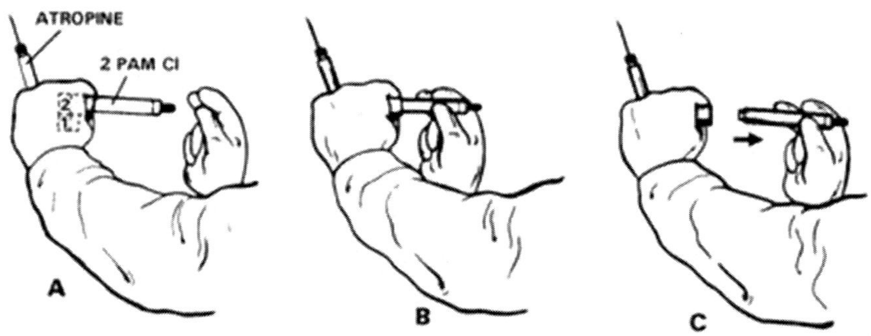

Figure 94. Proper method of holding the Mark One Injector following initial injection of atropine.

Perhaps the more recognizable exposure to a nerve agent in aerosol or vapor form is immediate eye pain with dimness of vision. Rhinorrhea (nasal secretions) and dyspnea (shortness of breath) are other signs. These symptoms are seen with oleoresin capsicum (OC) exposure. However, the pupils of the eyes in nerve agent exposure are pinpoint pupils, which will not be found in OC exposure. If a liquid droplet of nerve agent is on the skin, the muscles in this area may have fasciculation (twitching), visual difficulties, shortness of breath, and rhinorrhea may be delayed. Exposure to 70 mg/m3 of sarin will cause death in minutes.

The acronyms SLUDGEM, DUMBELS, or OBSERVANT can be used to describe nerve agent exposure:

- **S**alivation
- **L**acrimation
- **U**rination
- **D**efecation
- **G**astrointestinal upset, cramping, and nausea
- **E**mesis or vomiting
- **M**iosis or pinpoint pupils

Or

- **D**iaphoresis (sweating), diarrhea
- **U**rination
- **M**iosis (pinpoint pupils)
- **B**ronchospasm, bradycardia (slow heart rate)
- **E**mesis (vomiting)
- **L**acrimation (tears)
- **S**alivation

Or

- **O**thers affected suddenly
- **B**reathing, short of breath, difficulty breathing, chest tightness
- **S**ecretions, perspiration, coughing up secretions, drooling
- **E**yes, dimness of vision, eye pain, pinpoint pupils
- **R**hinorrhea, running nose
- **V**oiding, urination and defecation
- **A**rrhythmia, very slow, or very fast heart rate
- **N**ausea, abdominal cramps, and vomiting
- **T**witching, muscle twitching or seizures

If a nerve agent attack is suspected, evacuate upwind and keep persons possibly contaminated downwind of responders, the command post, and staging areas. Responders need to be in at least Level C attire. Physical contact should be avoided until the persons are decontaminated. Low-lying areas should be avoided and those entering the warm or hot zone should do so from upwind. EMS treatment should not be attempted, other than administration of the Mark One or DuoDote® medications through clothing. Decontamination will be a priority to expedite care. Decontamination will be clothing removal and washing with soap and water. Equipment can be decontaminated by 5% hypochlorite solution. If feasible, double bag clothing to prevent off gassing and further contamination of responders. Patients exposed to vapor without substantial symptoms and signs can be observed for one hour and released while those with substantial signs or symptoms or those with suspected liquid contact need to be monitored for at least 18 hours. Those with rhinorrhea (nasal secretions) and miosis (constricted pupils) who have not improved in one hour need a Mark One or DuoDote® administered.

Patients with sweating and fasciculation (muscle twitching) will need at least one Mark One or DuoDote® administered and transport to a hospital. An additional Mark One or DuoDote® may be required if symptoms and signs do not resolve quickly. Patients with continuing dyspnea, vomiting, and muscle weakness will need one or two Mark One or DuoDote®, and may require an additional kit. These require prompt transport to the hospital. If the person has collapsed or is seizing, they may require three Mark One or DuoDote®. Prompt transport to the hospital is critical. If the agency issues the CANA injector (diazepam), with seizing, one should be administered. Remember patients must be decontaminated prior to transport to a hospital. A substantial issue in the Tokyo subway attacks was the arrival of most of the patients on foot or by private vehicle. This meant the patients were not decontaminated and hospital personnel were not aware of the type of incident. The patients' clothing off gassed nerve agent affecting hospital personnel and the emergency facilities also became contaminated with nerve agent.

Vesicant Agents

These agents were developed in the 1800s by Germany. These agents produced more chemical casualties in WWI than any other agent. The Germans first used them in July 1917 in artillery shells against the British. During WWII, German and Italian bombing sank a US ship in an Italian harbor and caused the release of US vesicant agents. Records are sketchy, but some claim at least 600 US soldiers were exposed and as many as 100 died from vesicant agent exposure. Vesicants have been used repeatedly by a variety of countries even into the 1990s. It appears the US concealed the cause of the injuries from medical staff initially. The civilian casualties are not available from reliable sources. Blister agents may not have effects for several hours and are usually not lethal but disabling instead. Blister agents attack the exposed skin, eyes, and lungs. Their effects may be not felt immediately but do, in some cases, kill. Their effects are dose dependent. They produce reddened areas and blisters, thus the name blister agent. These agents tend to be persistent and present as an oily liquid.

These agents are designed as a denial of use or denial of area. These agents have minimal symptoms during initial exposure. Within 24 hours serious blisters and other symptoms occur. Examples of vesicant or blister agents are sulfur mustard (H), distilled mustard (HD), nitrogen mustard (HN1, HN2, HN3), and Lewisite (L).

The agents react rapidly and bind to tissue quickly. No early indications exist such that those exposed are contaminated and unknowingly spread the contamination. These agents tend to be oily and may have the odor of mustard or garlic. The areas affected include the eyes, airway, and skin, but absorbed agent can affect the gastrointestinal system and liver. The immediate effects are reddening of the skin (erythema), which may include itching, stinging, or burning. This may resemble sunburn or a first-degree burn injury. These may appear within hours of exposure or be delayed up to 24 hours. Other symptoms include hoarseness, sore throat, cough, dyspnea (difficulty breathing), and possibly nausea coupled with vomiting.

The eyes are very sensitive to vesicant agents, but long-term damage is not common. The skin develops large blisters and secondary infections are common. The damage to the airway is necrosis of the mucous membrane with muscle damage following. Lung damage can be profound and long term. Respiratory failure is possible and is the cause of death in vesicant agent exposure. The eyes may have miosis, which can be mistaken for nerve agent exposure, but one must remember the signs and symptoms of nerve agent exposure develop rapidly while vesicant agent signs and symptoms are slow to develop. The gastrointestinal system may be affected. The effect on the central nervous system is not well understood but seizures have been reported because of exposure (see Figure 95).

Response to a suspected vesicant agent attack requires responders to don at least Level C protective clothing. Respiratory protection is crucial in this setting. Vapors will likely collect in low-lying areas and those exposed may be contaminated heavily. Soap and water are used

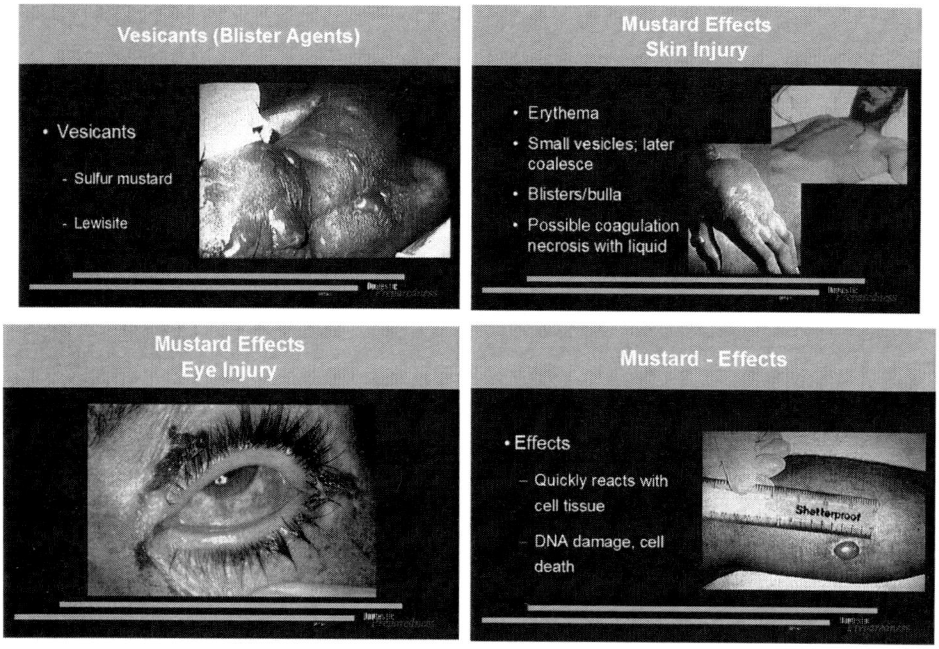

Figure 95. Vesicant exposure to the skin and eyes.

for decontamination. No bleach or HTH should be used for decontamination of equipment as it may react violently with vesicant agents generating heat and chlorine gas. Suspected contaminated clothing should be double bagged. Medical treatment is supportive.

Other vesicant agents include phosgene oxime and Lewisite, which induce immediate pain in the eyes and mucous membranes. This causes those exposed to retreat and wash the agents off with soap and water. This defeats the purpose of slow detection of exposure and prevents some degree of injury. The later symptoms are similar to other vesicant agents. Vesicant agents may be detected by electronic means, M8 or M9 paper, hazardous materials test kits, and military detection kits. If suspected the NAERG Guide 153 should be followed. The potential for cross contamination is very high. These agents can be synthesized in a "homemade laboratory" with moderate to very difficult levels using thiodyglycol as one precursor. Several methods and precursors can be used to manufacture vesicant agents. Mustard agent shells are still found buried and at sea. They may be trawled up as one was in 2010 near New York. The shell appeared to be a WWI mustard shell. The crew of the boat were exposed and had to be hospitalized.

Table 5. Toxicity Data

Agent	*Effect*	*Ct50 (mg-min/m^3)*	*Liquid on Skin*
HD - Sulfur Mustard	eye	12-200	
	pulmonary	100-200	
	erythema	200-1,000	10 micrograms
	death	1,500 inhalation	10,000 skin
L - Lewisite	erythema	>1,500	10-15 micrograms
	death	1,500 inhalation	40-50 mg/kg
CX - phosgene oxime	eye	200	
	erythema	2,500	
	death	3,200	

It appears that chemical weapons were haphazardly buried following WWI and WWII around and on military facilities. Some old military impact areas likely contain chemical agent artillery or mortar rounds, which failed to function. These shells and containers of chemical agent are still active and dangerous. Thousands of military chemical agent detection kits were distributed in the US during WWII. Most have not been located. These kits contain live chemical agents. It takes experienced military EOD personnel to identify these shells. The lethal dose in 50% of those exposed to vesicant agents usually is in the more than five grams dosage (see Figures 96 & 97).

Locations Chemical Weapons Have Been Found

- Dover Air Force Base, Maryland (blister agent, one injury)
- Long Island, New York (blister agent, one injury)
- Mobile, Alabama (phosgene)
- UN New York City (nerve agent)
- Washington, DC (1,000 ordnance items, assorted types)

Blood Agents

Blood agents, which will likely be cyanide-based, attack the body's metabolism at the cellular level. If one survives the exposure, no long-term effects usually occur, unlike blister and nerve agents. The issue with cyanide and blood agents is they are ineffective in open air and

Figure 96. Poster for identifying chemical agent testing kits.

require high doses to kill. These agents could be more effectively used in a closed setting such as a theater or building. Those who die will do so very quickly and will be dead before the type agent can be determined or antidotes administered.

Cyanide and blood agents attack the body at the cellular level. Agents such as hydrogen cyanide (AC), which is lighter than air, and cyanogen chloride (CK), which is heavier than air, and are usually dispersed as a gas. Cyanide salts may be in solid or dissolved in a liquid. Expedient hydrogen cyanide can be generated by exposing cyanide compounds to an acid. These agents require high concentrations to kill and are usually restricted to enclosed structures to be effective. The likely type device will be a metal container with a cyanide salt such as sodium or potassium cyanide combined with an acid. This will generate hydrogen cyanide gas. This could be fatal to those in proximity to the device. This is a similar method as used in the cyanide gas chamber.

When exposed to high levels of cyanide gas, victims collapse and may have seizures. An early sign of cyanide gas exposure may consist of sneezing, burning, or itching skin. Removal to fresh air may facilitate recovery with few ill effects. Those in serious distress will need a

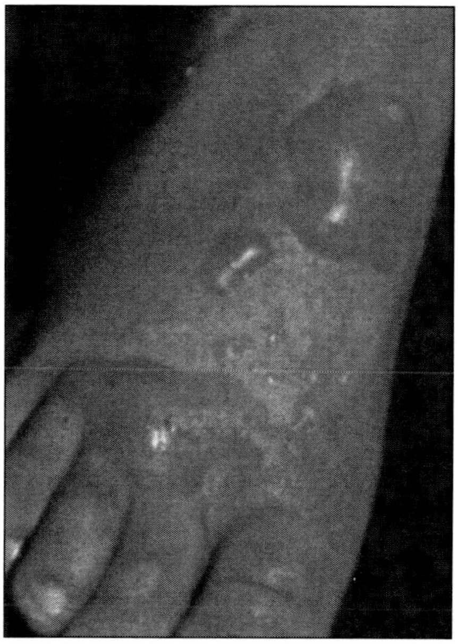

Figure 97. Foot exposed to blister agent.

cyanide antidote kit. These are not usually persistent agents if in vapor form. The key to differentiating this exposure from a nerve agent is the lack of pinpoint pupils. Cyanide is detectable using standard hazardous material instruments and military test kits. If cyanide is suspected, follow NAERG Guide 119 for cyanogen chloride or NAERG Guide 117 for hydrogen cyanide. Level B PPE must be used, as some cyanide agents will penetrate air purifying respirator filters. Removal of clothing will remove most cyanide gases but when feasible, have those potentially contaminated shower with soap and water. Most field EMS agencies will not carry cyanide antidote kits. In many instances, even medical facilities will have only a few of the antidote kits. However, most of the victims who will die will be dead on scene. Threats to contaminate water systems are not usually credible as to contaminate a large water system would require hundreds to thousands of pounds of cyanide be added to the system. However, this does not prevent the contamination of a small water supply such as a bottled water dispenser or concealing a small amount of cyanide in any liquid dispenser. Further, cyanide could be mailed in powder form as an inefficient weapon. The contamination of consumer products with cyanide has been used for criminal purposes to commit murders. The author has investigated and arrested individuals who stole a large quantity of cyanide in an attempt to contaminate a public water supply. This threat was not credible because tons of cyanide would be needed to contaminate the public water supply but the concern was the contamination of food, water dispensers, or conversion to hydrogen cyanide gas with an acid in an enclosed facility. The suspects were arrested prior to the use of the cyanide and both were convicted. The stolen cyanide was recovered. Their motivation was not clear but did not appear to be terrorism related. The local hospitals, when contacted, had two cyanide kits to cover a 30-county catchment area with a population approaching 750,000.

Table 6. Hydrogen Cyanide Concentration Effects (mg/m^3)

300 lethal within seconds		
200 lethal after 10 minutes		
150 lethal after 30 minutes		
120-150 fatal within 30 to 60 minutes		
50-60 tolerated for up to one hour		
20-40 minimal symptoms		
Agent	*Effect*	*Ct50 (mg-min/m³)*
AC–hydrogen cyanide	death	2,500–5,000
CK–Cynaogen	death	11,000

Pulmonary Agents or Choking Agents

Choking and pulmonary agents attack the lungs and cause pulmonary edema and fluids in the lungs resulting in "dry land drowning." Common agents include chlorine, phosgene, phosphine, hydrochloric acid, and oxides of nitrogen, smoke from some plastics, incineration of fluorocarbon compounds such as Teflon® or Freon® type compounds producing perfluoroisobutylene (PFIB), and anhydrous ammonia. Hydrochloric acid and phosphine may be found in clandestine drug laboratories at sufficient levels to cause serious injuries. These agents tend to be heavier than air and concentrate in low-lying areas. High concentrations are required to produce death, although low concentrations can injure and disable. The more common military agent is phosgene, developed in 1812. Phosgene and chlorine were used in WWI. The first attack by the Germans against the Allies in April, 1915 is reported to have produced 5,000 dead and 10,000 injured. The issue is 160 tons of chlorine was required to produce these casualties, which is inefficient. However, the terror produced by this introduction of a new method of warfare was substantial. In December, 1915, the Germans attacked with phosgene in artillery shells, which induced panic, as the Allied troops were unprepared. The ensuing attack exposed more than 1,000 soliders to the gas and killed more than 100. The British used both chlorine and phosgene in 1916 against the Germans. Open source reporting alleges that Syria used chlorine in 2014 in the current conflict. Even in the Civil War, both Union and Confederate forces considered using chemical weapons. Chlorine and cyanide were among the suggested substances. No data definitively confirms either side used chemical weapons during the Civil War.

These agents kill by impairing lung function and filling the lungs with fluid. Toxic industrial chemicals such as chlorine, anhydrous ammonia, or phosgene can be used. Other agents have been weaponized and include phosgene (CG), and diphosgene (DP). These agents produce a strong alkali when mixed with the water in mucous membranes and the lungs. These agents also injure the eyes. These agents will pool in low areas and displace oxygen, creating another hazard.

Chlorine has a pungent and distinct odor of bleach. It reacts with the water in eyes, mucous membranes, and lungs to produce hydrochloric acid. Exposure can produce death. Follow NAERG Guide 124. Anhydrous ammonia is used in farming and refrigeration systems. It can be used in the manufacture of methamphetamine. It has a strong ammonia odor and attacks the eyes, lungs, and any areas of the body, which are damp by forming ammonium hydroxide, a strong alkali. Follow NAERG Guide 125. Phosgene has the odor of newly mown hay but requires high concentrations to kill. It can be produced in fires involving fluorocarbons. Follow NAERG Guide 125.

Treatment for these exposures includes donning chemical protective clothing and at least Level B protection with self-contained breathing apparatus. Removal of those exposed to fresh air and out of low-lying areas is essential. Follow-up care includes decontamination by removing clothing and when feasible, washing with soap and water. Administration of 100% oxygen and placing the victim at rest is important.

Irritant or Riot Control Agents

Irritant agents tend to be less lethal and are used in many facets of law enforcement and by the military. These include oleoresin capsicum (OC) or pepper spray, CS (2-chlorobenzalmalononitrile) or tear gas, or the older less effective Mace or CN (chloroacetophenone). These agents attack the eyes and make breathing difficult but are not usually lethal. Those with asthma and pre-existing respiratory conditions may have these conditions exacerbated infrequently including fatal results.

Table 7. Toxicity Data

Agent	*Effect*	*Ct50 (mg-min/m³)*
CN - Mace	irritation	10-20
	death	14,000
CS - tear gas	irritation	5-10
	death	>50,000

These agents are typically considered not hazardous but can kill in high concentrations. Some individuals are allergic to MACE (CN). Some OC agents are dispersed using an alcohol base, which is flammable. OC mimics nerve agent exposure as it causes eye pain, nasal secretions (rhinorrhea), shortness of breath, and can cause nausea. These are similar to nerve agent exposure signs and symptoms.

Several events have been reported in which these agents have been used to force the evacuation of structures. The motives have ranged from retaliation to juvenile pranks. In one event, more than 100 people had to be treated and decontaminated for OC exposure after OC was delivered into the HVAC intake at a mall. Several individuals with pre-existing chronic obstructive pulmonary disease or asthma had to be transported to a hospital for emergency treatment and admission. In an incident involving a college dormitory, a police officer along with several students required emergency treatment in a hospital setting for the results of exposure to OC.

Decontamination includes removing clothing and washing with soap and water. These individuals will "off gas" or spread the fine crystalline particles of the irritants from their clothing and hair if careful decontamination is not performed. Those contaminated should be cautioned not to rub the affected areas. The initial application of water may intensify the effects for a short time. Those involved in these scenes and decontamination should wear at least Level C PPE. Medical treatment is supportive with bronchodilators and oxygen; rarely steroids may be required to reduce inflammation.

Exotic Agents

BZ or 3-quinuclidinyl benzilate is a hallucinogen with anticholinergic effects, which can cause delirium, dry skin, amnesia, and tachycardia (rapid heart rate). BZ usually is delivered as a fine powder. The individuals exposed to BZ are susceptible to warm environments and can experience heat stroke. Cooling is indicated for those with high body temperatures. Cardiac dysthymias are a common cause of death from BZ exposure. This agent is toxic and the complication rate is high. Individuals exposed to BZ should be decontaminated with soap and water. Clothing must be removed and rescuers must wear at least Level C protection. Reports exist that suggest BZ, an aerosolized fentanyl analog, mixed with halothane was used in the Russian Theatre hostage situation. It should be noted that a large number of the hostages died from the effects of the unknown agent or agents.

Adamsite (DM) is a vomit agent in powder form developed by the Germans in WWI. DM passed through gas mask filters of the time. DM was paired with choking agents necessitating removal of the air-purifying respirator to clear the vomit allowing inhalation of the choking agent. DM causes persistent vomiting and is a mild irritant. Level B PPE is mandated since DM may penetrate air purifying respirator filters. Clothing should be removed and decontamination is soap and water. Treatment is supportive with antiemetic drugs.

Smoke

Smoke agents are used for obscuring, but most agents cause irritation of the eyes and respiratory tract. In liquid form, smoke agents may cause burns to skin while some smoke grenades produce acid gases similar to pulmonary agents. White phosphorus grenades produce toxic smoke and the burning white phosphorus will adhere to skin. White phosphorus is hard to extinguish as it burns whenever oxygen is available. These wounds may have to be covered with a wet dressing or submerged to stop the burning. A 2% to 4% copper sulfate solution is used to inactive burning white phosphorus. Any white phosphorus must be removed from skin or any embedded must be removed as it generates toxic compounds, which are absorbed by the body. Many of the smoke-producing devices produce smoke by a chemical reaction or burning which produces heat. Do not handle expended smoke grenades or approach them if they are still producing smoke. Eye irritation should be treated by flushing with water or saline. Respiratory difficulty should be treated with fresh air, oxygen, and bronchodilators.

Detection of Chemical Agents

Detection of chemical agents can be accomplished using sensors with a photoionization detector. This type instrument will detect levels of oxygen, the presence and level of explosive gases, chemical warfare agents, toxic industrial chemicals, halogen gases such as chlorine or bromine. Many may detect volatile organic compounds, pepper spray, and carbon monoxide. These instruments require calibration and testing periodically. Other detectors may use a color change to indicate exposure to an agent or chemical. M8 and M9 kits, along with the M256 kits are examples as is the tube type colorimetric detectors. Cross-reactions are common as are false positives with colorimetric detectors.

Biological Agents

Biological agents have been in use for many years. Some texts report 600 BCE that Assyrians used ergot from moldy rye to poison wells of their enemies. Ergot is a hallucinogenic agent, which causes bizarre and frequently violent behavior. During the siege of Kaffa in 1346, bodies of plague victims were catapulted into the walled fortress in hopes of infecting the defenders. Japanese Unit 731 conducted extensive biological warfare research and used biological weapons against the Chinese during WWII. Human-to-human transmission can occur with some diseases such as influenza, Ebola, or smallpox. One study postulated a single person spending a day in a busy airport might generate some 40,000 smallpox cases within a week. However, direct delivery methods are not usually efficient. The salad bar contamination with salmonella by the Rajneeshee in Oregon in 1984 sickened more than 750 persons but only hospitalized roughly 45. The anthrax attacks of 2001 only managed to infect roughly 20 persons and to kill five individuals. However, the fear produced by these attacks was far out of proportion to the threat produced by the attacks.

Biological agents may be introduced into communities via vectors such as fleas for plague; bats are suspected of transmitting SARS, MERS, and Ebola. Mosquitos transmit yellow fever, VEE, malaria, and West Nile virus. Fomites on clothing, blankets, or pillows can transmit smallpox. Biological agents consist of living bacteria or rickettsia which are disease causing organisms, their spores, viruses (many classify viruses as nonliving), and toxins derived from living organisms. The virulence of the agent depends upon its type and method of attack. For instance, anthrax in the skin or gastrointestinal form is dangerous but not deadly in most cases. If inhaled and not treated, the pneumonic version is usually fatal. The period from exposure to anthrax and symptoms appearing ranges from one to 60 days, but usually about a week from exposure passes before symptoms appear.

In many instances in the cases of bacteria and rickettsia, antibiotics are usually effective if started early. Weaponized versions of these diseases may be antibiotic resistant. Few antiviral drugs are available, making viruses a good weapon. One feared viral weapon would be smallpox, as most of the population has not received a vaccination or if they did, no longer has sufficient titers of antibodies to resist this virus. Supposedly, the only sources of this virus exist at the Centers for Disease Control and Prevention in the US and a similar facility in Russia. However, the finding of an unknown smallpox supply in a laboratory in the US has cast into doubt the elimination of smallpox as to availability. The last known case of smallpox occurred during the 1970s.

Another concern is the communicability of the disease. While deadly, anthrax is not communicable in normal circumstances, whereas the deadly Ebola virus, at least in the airborne variants, is said to be easily transmitted. The pneumonic form of the plague may be transmissible via the airborne particle route.

Toxins derived from living sources include items such as ricin, botulinum toxin, tetradotoxin, and snake venom. The production of ricin is simple and it has been used in several incidents including at least one producing death. As little as 4,000 mg of Sarin nerve agent on the skin is lethal while ricin is roughly 10,000 times more toxic. Comparative Lethality of Toxins and Chemical Agents in Mice

Table 8. LD50 (g/kg) Source

Botulinum toxin	0.001	Bacterium
Shiga toxin	0.002	Bacterium
Tetanus toxin	0.002	Bacterium
Abrin	0.040	Rosary Pea
Diphtheria toxin	0.100	Bacterium
Maitoxin	0.100	Marine Dinoflagellate
Palytoxin	0.150	Marine Soft Coral
Ciguatoxin	0.400	Marine Dinoflagellate
Textilotoxin	0.600	Elapid Snake
Clostridium perfringens	5.000	Bacterium
Batrachotoxin	2.000	Poison Frog
Ricin	3.000	Castor Bean
alpha-Conotoxin	5.000	Cone Snail
Taipoxin	5.000	Elapid Snake
Tetrodotoxin	8.000	Puffer Fish
alpha-Tityustoxin	9.000	Scorpion
Saxitoxin	10.000	Marine Dinoflagellate
VX	15.000	Chemical Agent
SEB	27.000	Bacterium
Anatoxin-A	50.000	Blue-Green Algae
Microcystin	50.000	Blue-Green Algae
Soman (GD)	64.000	Chemical Agent
Sarin (GB)	100.000	Chemical Agent
Aconitine	100.000	Monkshood
T - 2 Toxin	466.000	Fungal Mycotoxin

Dissemination of biological agents will determine how effective they are in terms of infection. Although explosives can be used to distribute biological agents, the thermal event will destroy many of the agents. Other methods to distribute these agents include: simply mailing them in a powder form as occurred in the anthrax attacks in the US and several of the ricin attacks. They can be distributed in a powder form, inserted into food as occurred in the attacks in Oregon, or on objects such as the smallpox transmission to Native Americans in 1763 from contaminated blankets (although this has been disputed as a legitimate transmission method). Other methods might include usage of agricultural sprayers including aircraft. In the Oregon salmonella attacks, hand sprayers were reportedly used to contaminate salad bars.

The ideal size of the biological agent to be inhaled is between two and 10 microns with five microns optimal. This is difficult to achieve in weaponizing biological agents. The success of the US anthrax attacks were in part because the appropriate size of the anthrax spores that allowed them to be inhaled with casual dispersal.

To complicate matters, some of the biological agents occur naturally such as anthrax and plague, making differential diagnosis difficult. For instance, plague is endemic in the southwestern US such that cases occurring in New York City with persons with no history of travel would likely indicate a biological attack. However, anthrax transmission has occurred with imported animal hides. The recent scandals involving the CDC demonstrate that security of the biological agents such as anthrax is not as secure as once believed. Further, since many of the agents are present in the environment as natural agents, the acquisition may not be difficult. The use of salmonella, supposedly obtained from chicken, used in the biological attacks in Oregon,demonstrates the ease with which naturally occurring bacteria can be used as a biological agent.

Indicators of biological agent usage may include threats. These may be verbal, written, or electronic. Intelligence may indicate a group has the capability to deploy a biological agent and the location is within the group's target spectrum. Other physical indicators may be sprayers, tubing, pressurized gas tanks, Petri dishes, culture or growth media, or similar items on site or within a laboratory setting. Containers labeled biohazard may also provide a sign that a biological agent may have been deployed or is present.

Discovery of a biological attack is usually after the fact with incubation or latent period prior to symptoms ranging from hours to days. Numerous persons developing similar symptoms with a common exposure such as working in the same building, eating in the same restaurant, shopping in the same mall or similar factors may indicate a biological attack. The confusing factors that include most biological agents produce "flu-like illness" as the presenting complaint. This may obscure the true nature of the disease or exposure. Further, many of the agents occur naturally, making the determination of an attack difficult to different from a natural outbreak. Respiratory distress is another common finding of many of the biological agents further complicating diagnosis.

Unknown Powders

When unknown powders are present, the primary effort should be to isolate and prevent the spread. If an item is labeled with "anthrax or ricin," and has no visible contamination, it should be double bagged. Those handling the time should immediately wash their hands with soap and water. The areas potentially contaminated should be isolated and law enforcement notified.

If an item has been opened and powder is visible, the item should be placed such that contamination spread will be limited. This may involve placing the item in a trash can or leaving it in place. If possible, the item should be covered without disturbing it. The individuals and area should be isolated. Those exposed should at least wash their hands with soap and water. Any fans or other air circulation devices should be turned off. If a large amount of powder is present or if the powder is visible in the air, shut down the HVAC system. Notify local law enforcement. They will make a decision, usually in concert with the local hazardous materials unit, as to what steps to take regarding decontamination. Some agencies will test the item first while others will proactively decontaminate those exposed. Doffing clothing into plastic bags, which will remove most of the contamination, and a shower using soap and water with special attention to the hair, is usually sufficient.

Many items are used as hoaxes to simulate anthrax spores or ricin. However, one cannot rule out other toxic materials such as insecticides, rodenticides, or radioactive material. Common hoax powders include talcum, chalk dust, flour, sugar, inositol, cornstarch, baker's yeast, or finely shredded paper. Familiarity with the jurisdiction's unknown powder policy and action plan is essential. The primary effort is to limit contamination spread while determining the character of the substance. Secondary considerations are decontamination. Laboratory testing may be necessary to differentiate the material from a biological agent. One must remember the powder may be a radioactive material or toxic chemical such as a pesticide.

Anthrax

Anthrax, bacillus anthracis, is a hardy spore-forming bacteria that can cause several forms of infection. The spore may be dormant for years in soil. The more common forms of anthrax are cutaneous anthrax where the spore enters a break in the skin or is ingested to provide the gastrointestinal form. The routes have a low mortality rate. The more dangerous of the exposure routes is the inhalational route. As few as 100 spores, some authorities argue less than this number, may cause inhalational anthrax. If the treatment with antibiotics is not started prior to significant symptoms being present, the mortality rate can be very high. Treatment for exposure includes antibiotics, usually ciprofloxacin. Time of death from symptoms, shortness of breath and widening of the mediastinum on x-ray, onset for respiratory anthrax is usually about 24–48 hours. The mortality rate is above 90% according to the CDC.

Common forms of anthrax include the respiratory route, which is usually fatal if not treated prior to symptoms appearing, the skin, or gastrointestinal forms. The mortality from the latter two runs between 5% and 20% according to the CDC if left untreated. These occur from anthrax spore exposure to nonintact skin or eating anthrax contaminated food.

The difficulty with anthrax is not its collection or growth but the "milling" of the spores into an approximately five-micron size. This size is reputed to be optimal for inhalation. The danger with inhalable anthrax is that its signature may not be noted until lethal symptoms appear. The recent presumptive releases of anthrax spores in the CDC laboratory demonstrates the danger of weapon-grade anthrax as did the alleged releases of anthrax spores in Sverdlovsk in 1979. The Sverdlovsk release is alleged to have killed about 100 individuals living downwind of the anthrax production facility. The failure of maintenance personnel to install replacement HEPA filters was the cause of the release.

Onsite testing can be problematic for anthrax as many environmental tests provide a false positive. During the 2001 anthrax attacks in the US, the author responded to more than 250 credible anthrax threats and hoaxes. Some of the calls had substances that appeared consistent with anthrax or other biological agent and tested positive with field tests, which test for protein. For instance, baker's yeast will test positive with some field tests as will nondairy coffee creamer. When the item tests positive for protein, environmental samples must be sent to a laboratory rated, usually at the state level, to perform confirmatory tests. This is an expensive and time-consuming procedure.

This makes credible threats, with a credible substance that tests positive, be treated as an incident involving a biological agent. The ease of transmission of anthrax in the environment makes containment difficult. Each agency must have a clear and practiced protocol to isolate, contain, prevent the spread, and decontaminate those exposed to a presumptive biological agent. Respiratory protection in suspected events is critical.

Persons exposed to what is reasonably believed to be anthrax should have expedient decontamination started. Use of water, even a garden hose, is suggested, doffing clothing and washing with soap. If such is not available, the person exposed during clothing removal should wear the use of a N95 respirator. Responders should be in at least Level C attire preferably Level B PPE. Notification of EMS personnel and the receiving hospital is essential if those exposed will be transported to a medical facility. In many locales, the public health department may play a role in treatment of potential exposures and confirmatory testing.

Ricin

Several events in the US have involved the manufacture of ricin using castor beans. The manufacture of the product is not difficult. Ricin, even in crude form, is very toxic with some authorities suggesting that as little as two milligrams may be lethal. Ricin exposures may occur from inhalation, injection, or through nonintact skin. Efforts to use solvents such as DMSO to carry the ricin molecule through intact skin have not been successful because of the large size of the ricin molecule. However, exposure through non-intact skin is possible (see Figure 98).

Ricin usually appears as a white or off white crystalline material in powder form. One issue is a laboratory manufacturing biological agents is similar to one manufacturing methamphetamine or explosives and the finished products appear similar. Officers should be alert for castor beans or other biological material, Petri dishes, glass laboratory apparatus, and even an incubator, which may suggest the laboratory is a biological laboratory. Use of chemical protective clothing and respiratory protection is critical in this setting. Onsite testing is problematic with false positives common. Standard decontamination pro-

Figure 98. The castor bean plant.

cedures should be used. Responders should be in at least Level B PPE and use of Level A should be considered.

Medical treatment is supportive for ricin exposures. No definitive medical treatment exists. Symptoms, which include shortness of breath and flu-like illness, may begin within eight hours but may be as long as 12–24 hours. Pulmonary edema is a common finding as are necrotic lesions to liver and spleen followed by circulatory collapse and death. This agent has been deployed or manufactured by several domestic terrorist groups and individuals.

Infectious Agents

The plague caused by the bacterium yersina pestis is also known as the "Black Death." This disease is endemic to the US and presents as a flu-like illness. Flea bites usually transmit plague. The symptoms may include respiratory distress, bloody sputum, and fever, followed by circulator collapse and death. Tetracycline and other common antibiotics can be used to treat the disease in its natural form. This agent has been weaponized and speculation exists that an antibiotic resistant form of plague may exist as a biological agent.

Other common agents that have been weaponized include the bacteria tularemia and rickettsia "Q" fever. Both produce flu-like illnesses and are incapacitating but are not usually fatal. Q fever may produce myocarditis and meningitis. The treatment is common antibiotics.

Viral agents include smallpox, Venezuelan equine encephalitis (VEE), and viral hemorrhagic fevers. Viruses highjack the cell reproductive apparatus and force it to produce more viruses. This may lead to an explosive growth in the number viruses present. Antiviral agents are in their infancy and many have to be specific for a particular virus. This makes their use problematic. The better method of resisting viral diseases is through immunization to develop sufficient antibody titers to ward off such agents.

Smallpox has been used as a biological weapon and is easily transmitted. Although the only stocks of smallpox allegedly reside with the CDC and its Russian equivalent, the recent discovery of smallpox in a non-secure laboratory cast doubt as to its security. Smallpox is transmitted via aerosol. This is a very contagious disease with an incubation or latent period of seven to 17 days. One of the first signs is a pox-like rash with flu-like illness. Most individuals have not been vaccinated recently and likely do not have sufficient antibody titer to exhibit immunity to the disease. This agent has been weaponized and used as a biological agent.

Venezuelan equine encephalitis (VEE) is another viral agent that has likely been weaponized. This is a mildly contagious disease with an incubation or latent period of about one to four days. This disease presents with flu-like illness. This disease incapacitates but is usually not fatal.

Viral hemorrhagic fevers (VHFs) occur naturally and some are very contagious. The mortality rates can be very high on some of the VHFs. Speculation exists that these agents may have been weaponized. These present with flu-like illness. These include Lassa fever, Ebola, Crimean-Congo fever, Marburg virus as old world VHFs, but a new world VHF is Hantavirus. Other substantial viral illnesses, which appear to be of natural origin, include severe acute respiratory syndrome (SARS), Middle Eastern respiratory syndrome (MRE), and the various strains of influenza.

The author dealt with an atypical influenza outbreak that killed several young adults and attracted national attention. The event placed a substantial strain upon public safety and medical resources with those ill and the "worried well" seeking medical attention. Hospital emergency departments, walk in clinics, and physician offices were overwhelmed with those who experienced typical influenza symptoms but were not seriously ill. One can conceive of the massive strain, which would be produced, should a biological agent be used. The psychological effect might be more substantial than the disease itself.

Some public health authorities have postulated that during the influenza season, well-timed "terrorist communiques" claiming the release of a viral agent producing symptoms consistent with influenza could easily overwhelm medical resources throughout the country. The differential diagnosis of influenza-type illnesses is difficult during the early phases of the disease. The simple threat with no underlying effort would be easily accomplished.

Toxins

Toxins are materials derived from a variety of organisms, which include plants, snakes, sea life, and fungi. These toxins range from mild to extremely toxic. Snake venom is an example of a common toxin. A toxin encountered by law enforcement personnel is the botulinum toxin. This toxin occurs with the anaerobic reaction of materials used by inmates in an attempt to produce alcoholic beverages in jail. Outbreaks of inmates intoxicated by botulinum toxin happen each year.

The impact of this toxin is the suppression of muscle movement, which affects swallowing, and breathing. The effects are long-lasting. Some inmates exposed to this toxin spend lengthy periods on a respirator. Thus, the ease of manufacturing such a toxin can be seen. Vaccines are available but must be used early into the event to be effective. Besides paralysis, the symptoms include diplopia (double vision), ptosis (drooping eyelids), and generalized weakness. These toxins have been weaponized. The author is aware of two cases where individuals overdosed themselves with botulinum toxin by using the wrong strength toxin in an attempt to "reduce wrinkles." This led to a lengthy hospital stay on a respirator. Both survived.

Mycotoxins are toxins derived from fungi. Although not common, these have been weaponized and allegations of their deployment exist. They are sometimes called "yellow rain." Their effects include GI bleeding, skin inflammation, hemorrhage, and death.

Biological weapon delivery is not effective when dispersed by an explosion. The thermal event and physical damage caused by the transonic shock wave will render a substantial portion of the biological material inert. Therefore, dispersal will likely use another method. Common methods include using a spray bottle as was used in the Oregon salad bar biological agent deployment, dispersal via sprayers or from an aircraft, simple dispersion via a powder in an envelope. The key to successful dispersion will be related to particle size, five microns is optimal, as this will affect inhalation of the agent into the lungs.

Radiological Weapons

These are weapons that use radioactive material as a weapon. Radioactive material emits harmful particles or rays. The nucleus of an atom contains massive particles such as the proton, charged positively, and neutron with no charge. Orbiting the nucleus is a negatively charged particle called an electron. The ratio of neutrons to protons determines the stability of the nucleus. Unstable atoms emit radiation in an effort to return to a stable state. Unstable atoms have excess mass or energy and emit mass in the form or particles or energy. The energy is in the form of electromagnetic waves. These particles and waves come from the unstable nucleus of an atom. This is called radioactive decay. A term used in describing radioactive decay is half-life. This is the time it takes the amount of radiation emitted to decrease by 50%. The approximate half-lives of some radioactive materials are listed below:

Iodine 131 (I131)–eight minutes
Cobalt 60 (Co60)–five years
Cesium 137 (Cs137)–30 years
Plutonium 239 (Pu239)–24,000 years
Nickle 59 (Ni59)–76,000 years
Uranium 233 (U233)–159,000 years

Isotopes are variants of atoms that have a differing atomic number, the number of protons present along with neutrons within the nucleus. This may cause the nucleus to become unstable. The nucleus may emit particles or electromagnetic rays to return to a stable isotope. Not every isotope is unstable and not every isotope emits radiation. For instance, hydrogen in its normal form has one proton in its nucleus and is stable and not radioactive. A hydrogen isotope called deuterium has one proton and one neutron but is still stable and not radioactive. Tritium, another hydrogen isotope, has one proton and two neutrons. Tritium is radioactive and emits beta particles. When tritium and certain dyes, which become phosphorescent when exposed to beta particles, are embedded in glass tubes, these tubes glow. This is the tritium sights seen on firearms. The varying radioactive isotopes of elements may emit alpha particle, beta particles, gamma radiation, or a combination of these particles and rays. Isotopes may be radioactive but have the same chemical properties as the stable isotopes of the same element. In unusual circumstances, neutrons may be emitted. These circumstances are from specially constructed radiation sources or from active fission.

Radiation causes damage through a variety of methods. It causes damage to DNA, which may cause mutations, induce cancer, kill the cell, or cause cancer. Tissue that divides rapidly (grows) is more likely to be affected. This includes the lens of the eye, bone marrow producing blood cells, hair follicles, and the lining of the gastrointestinal tract. Radiation may also induce burns to the skin and tissue from the energy imparted to the skin and tissue as the radiation particles or waves strike tissue. Radioactive material inhaled or ingested presents a more serious hazard as the radioactive material is in direct contact with tissue. Many radioactive materials are toxic heavy metals such as uranium and plutonium. Some may accumulate in certain organs. Radioactive strontium 90 accumulates in bone; radioactive cesium 137 accumulates in the liver, while radioactive iodine 131 accumulates in the thyroid. In cases of radioactive iodine exposure, individuals may be instructed to take iodine supplements to flood the thyroid with iodine and prevent its uptake. Iodine 131 may be released when fuel bundles in a nuclear power facility are breached and the containment leaks. This type event occurred with the Three Mile Island nuclear incident and Chernobyl and Fukushima nuclear power plant explosions and fire.

Radiation in high doses kills cells directly while lower doses cause damage to the DNA that may cause mutations or cancer. Human eggs and sperm are damaged by radiation and may cause congenital deformities in children. Radiation suppresses the white blood cell levels and many exposed to high levels of radiation die of infections.

Alpha particles are emitted from the center of the atom called the nucleus. Massive particles called protons, with a positive charge along with neutrons with a neutral charge, reside there. An alpha particle consists of two protons and two neutrons. Alpha particles have a short range and clothing will stop their radiation. However, if inhaled or ingested, they produce tremendous damage internally since no shielding is present to protect the tissue exposed.

Beta particles are emitted from the nucleus and have the mass and charge of an electron. Electrons are particles with a negative charge, which orbit the nucleus. Beta particles also can be stopped by clothing and they, too, if inhaled or ingested, can cause tremendous internal damage, as no shielding is present to protect tissue. They can cause skin damage, as they will penetrate a few millimeters in tissue. They have a short range in air.

A rarer type of radiation is neutron radiation. Neutrons are the only form of radiation that might interact with objects, or organisms exposed and induce atoms within the structure to become radioactive as the neutron is absorbed. This changes the neutron to proton ratio and may cause the nucleus to become unstable. The unstable nucleus will emit radiation. Neutrons are usually only seen in a fission-type environment. Thermal neutrons are associated with nuclear reactors and nuclear weapons. However, some radiation sources can produce neutron radiation. These usually contain an alpha-emitting radioactive material surrounded by beryllium. These devices are used in many instances to measure density of substances such as earth. Construction firms, departments of transportation, and paving companies are likely locations of this type device.

Gamma rays are electromagnetic radiation, which requires substantial shielding such as lead, concrete, or soil to stop. These rays can cause damage if the exposure is external and if radioactive material is ingested or inhaled. The external exposure goes away when the person moves away from the source, but the internal exposure continues until the radioactive material is excreted or decays.

Radiation is measures in several units. The older measures, which many instruments still report, are Roentgens (R). Roentgens are a measure of radiation ionizing effects in air. RADs are Radiation Absorbed Dose and measure the amount of energy deposited in tissue. REM, Roentgen Equivalent in Man, defines the damage performed by the radiation in humans and is adjusted based upon the type radiation. For practical purposes, they are the same. As a rule, gamma radiation and neutron radiation cause more damage, in most cases, than do alpha or beta particles. Neutrons are more dangerous because if captured by a stable isotope, the neutron may cause the atom to become radioactive. The emission of radiation can be measured in activity of a sample. The older measurement is disintegrations per second (DPS) or Curie. The Curie (Ci) is 3.7×10^{10} DPS. The new measurement is Becquerel, which a Curie would be 37GBq (37 Giga-Becquerel). Newer measurement also includes the Gray and Sievert.

- 1 Gy (Gray) = 100 rad
- 1 mGy (milliGray) = 100 millirad
- 1 Sv (Sievert) = 100 rem
- 1 mSv (milliSievert) = 100 mrem
- 1 Bq (Becquerel) = 1 disintegration per second (DPS)
- 1 Curie (Ci) = 3.7×10^{10} DPS or 37GBq

Individuals have a large variation in their reaction to radiation doses such that the effects vary. The allowable occupational exposure to radiation is five REMs per year. During an emergency, a rescuer can receive 25 REMs with little short-term effect other than changes to the white cell count. This is upper end of emergency exposures that most agencies allow. However, exposures to as little as five REMs in a short period may cause detectable changes in white cell counts. During an emergency, the exposure to levels, which would exceed the occupational maximum of five REMs per year, should be voluntary. Further, older males should be used in these instances as those of childbearing age, both male and female, will likely incur more risks of long-term damage and potential for congenital defects in any children conceived. The cancer rates of the older worker would be lower because they are further into their expected lifespan. Exposures in the 50 to 100 REM dosage range may cause radiation syndrome, but such is usually mild, limited to nausea, fatigue, and hair loss. Blood cell production will be decreased. Exposures to 400 REM dosage will cause some persons to die with exposures above this limit making death more probable and survival times less. The initial organ affected most is the bone marrow, which will produce fewer white and red blood cells. This is seen above five to 10 REM exposures. At roughly 400 REM dosage, the organ affected next is the lining of the gastrointestinal system, which will die. This leads to infection and blood loss. At several thousand REM dosages, the last syndrome occurs. This is the central nervous syndrome involving the destruction of nerves and brain tissue. Death will occur quickly with this level of exposure.

Most agencies have specific dosage limits as shown by direct reading dosimeters. These may be pencil dosimeters or electronic dosimeters. One issue to remember is that these devices may have a variability rate of up to 20%. With this factor in consideration, many agencies set a dosage limit and then reduce it by half. This is to facilitate minimizing exposure including the dosage received during exit from the scene and the fact that the instruments are accurate plus ±20%. Some agencies have turn-back points such that when the radiation level reaches a certain point, they proceed no further into the scene. In nuclear power emergencies, this turn-back point in many locales is 100 mrem/hr for general activities. This may be increased if the activities involved are critical or involve lifesaving. Some agencies allow up to 200 REM dosage for lifesaving while others limit such to 25 REM dosage, while others limit such to five REM dosage. Significant variability exists among agencies in terms of emergency dosage allowed for lifesaving activities. One issue is if the person performing the rescue will receive close to 200 REM dosage during the rescue, the victim likely already has received a fatal dose. Risk to rescuers must be balanced with the probable survivability of the person to be rescued.

Explosions or other means may disperse radioactive materials. Such a device is a radiological dispersion device. The radioactive material is usually resistant to the physical damage of an explosion. Other methods such as manual delivery, dispersal from an aircraft, or even placing radioactive powder in an envelope via the US Mail are feasible. In most cases, the radioactive material will not kill or seriously injure most exposed in the short term but will cause extensive economic damage. Area denial is one use of this process. Further, decontamination is expensive, resource intensive, and time consuming. The areas contaminated must be decontaminated to prevent long-term exposure, which will likely cause negative health effects, and to prevent ingestion or inhalation of the materials.

Another type of deployment might be a radiation exposure device. This device could be concealed in a public place and those frequenting or passing by would be exposed to radiation. This type of device would induce substantial concern for those who had frequented the area and would raise concerns that similar devices might be emplaced at other locales. This would cause terror out of proportion to the threat posed.

Some public safety officials have postulated that terrorists might use an explosion to lure first responders into a zone contaminated by radioactive material. The primary issue is that this is not an immediately lethal weapon, but exposure can have long-term health issues. The only method to detect radioactive material and radiation is a radiation detector. This means that responders must be alert for issues related to explosions. The use of radiation detection "pagers" and use of scintillation instruments is recommended at the scene of any explosion, which appears to be a criminal act.

Decontamination is costly and in many instances, the items contaminated are buried in a radioactive material landfill, as the cost to decontaminate the item is more expensive than the item is worth.

Most insurance policies do not cover radioactive contamination. This may produce enormous economic losses. The loss of access to the area may further enhance the economic impact of such an event. The public will react in a negative fashion as the dangers presented by radiation have been overblown by the media while the primary function of such a weapon is both terror and economic.

Nuclear Threats

The threat of use of nuclear or radiological dispersion weapons (RDW) against the US by terrorists is not a new threat. There have been threats to use nuclear bombs and radiological devices against the US since the 1970s. In 2012 saw threats against the Queen of England in London and threats against New York City. Open source materials show more than 100 threats during 2014. Threats have been made in 2011 and 2012 against Washington, DC, against airliners in 2015, against Chicago Amtrak in 2004, Los Angeles in 1975, and Boston and New York City in 1976. Open source materials show more than 20 deployments of the nuclear emergency search team (NEST) in 2001. In 1970, the city of Orlando, Florida was a victim of nuclear extortion and received a workable plan for a nuclear weapon. The device was to be detonated if money was not received. A juvenile who did not have such a device committed the crime. Other cities experiencing extortion include Boston, Des Moines, and Lincoln, Nebraska. Plans for nuclear bombs are public record. The essential components special to nuclear material are probably not obtainable.

What is possible is the theft or purchase of a radioactive material. This could be dispersed by explosion or used for extortion purposes. Several hoax devices have been located by law enforcement that are purposely designed to resemble nuclear weapons. For even qualified personnel, these devices were realistic in appearance as the superficially resembled nuclear weapons of the man portable size. What is most probable is the use of a radiological dispersion weapon (RDW) or radiological dispersion device (RDD) that disperses either a long-lived isotope contaminating a wide area, or releases of a biologically active isotope that affects those exposed more rapidly. Long-lived isotopes can contaminate and render an area uninhabitable for many years, while biologically active isotopes, such as iodine, attack specific organs (thyroid) or systems in the human body causing serious health effects such as cancer. These isotopes are usually not only radiation hazards but are also poisonous as many are heavy metals.

There have been recoveries of large quantities of such isotopes recently that are suitable for radiological weapons. Isotopes that lend themselves to such a device are plutonium, cobalt-60, cesium-137, iridium, and others. Plutonium is probably more difficult to obtain, but the most significant radiation exposures outside Chernobyl have been incidents involving Cobalt-60 and Cesium-137 (Cs-137). Both of these radioactive isotopes have long half-lives (30 years for Cs-137) and are energetic. Both these isotopes are used for industrial radiography where their strong gamma rays penetrate metals to detect imperfections. They are also used in the sterilization process of medical materials, cancer treatment, and food processing.

One incident occurred in Mexico in which the Cobalt-60 source was removed from an abandoned industrial radiography system and placed in a home. The residents became critically ill and several died because of the high doses of radiation. The other event occurred in 1987 in Goiania, Brazil. A cancer teletherapy radiation source (Cs-137) was stolen and broken open. Extensive contamination and radiation exposure took place. This incident is second only to Chernobyl in terms of radiation exposure and contamination according to some authorities. More than 250 persons were contaminated and irradiated with more than 60 requiring hospitalization and several dying. A substantial portion of the city was contaminated. A significant increase in leukemia has been detected in this locale. Some authorities estimated the number of persons contaminated in the thousands.

The more likely scenario remains the theft of or other method to acquire a long-lived energetic isotope with dispersion via an explosive device. The isotope could be dispersed via airdrop or dispersed via liquid or aerosol. An extensive area could be rendered uninhabitable and serious health consequences would be experienced by those exposed not only from the external radiation but also through absorption by inhalation or ingestion. This would internalize the radioactive material increasing the radiation dose by a radioactive body burden. This radioactive material body burden will be exposing the chemically toxic isotope to the body.

Minimal security exists to guard radiographic sources, and most incidents involving theft, loss, or overexposure from radioactive material in the US is from this type of source. Simple industrial security and in many instances no security exists other than locks to guard this radioactive material. Theft of radioactive material or purchase on the open market is simple and inexpensive. If the radioactive material is not stolen it might obtained from shipments of radioactive materials, radioactive waste shipments, or storage facilities. Detection of radioactive material is by radiation detectors. Older Civil Defense type units can be utilized with energetic gamma-beta emitters such as Cs-137. Iridium-192, or Co-60, however, biologically active and medical isotopes such as radioactive iodine are not sufficiently energetic to be detected in many cases by this equipment. Specialized training and equipment such as gamma scintillation detectors may be required. Identify these resources prior to need and in most instances, they are available at the state level.

A "dirty bomb" can be used for terrorism in several scenarios, of which the most likely is a radiological dispersion device (RDD). This is a device designed to disperse radioactive materials through either an explosion or more simple means such as spilling the radioactive material into a ventilation system. Some devices have simply been radioactive material mailed to the target. Any explosion or unknown powder should be checked to determine that radiation and radioactive material are not present. FBI certified bomb squads, hazardous materials units, and state radiological health units will have radiation detection instruments. These devices are denial of use devices and the radioactive material may present long-term health issues for those exposed.

An RDD presents a threat where radioactive material is dispersed by an explosion. This presents the typical bomb hazards along with radiation exposure. An RDD with a sizeable amount of radioactive material could contaminate a large area. This necessitates an upwind approach to an explosion site when feasible. If the agency experiences such an event, the agency should notify its chain of command, state radiological health, and FBI. The agencies likely will have additional responders from NEST and military support units. Because of the long-lived radiation issues, this hazard will not be resolved quickly like many other hazardous material events. Another threat will be the nature of the radioactive materials as many are toxic chemically in addition to being radioactive. If a medical isotope is used, the radioactive material may be biologically active and bio-concentrate in an organ such as the liver or thyroid.

The critical factor is determining that a radiation hazard exists is using radiation detection equipment to do this. This knowledge will allow personnel to don protective clothing and respiratory protection. Time, distance, and shielding work in radiation exposure to minimize injury as with one minimizing time in the exposure area, staying as far away as possible, and when feasible, keeping solid structures between him or her and the radiation source. For radiation emergencies, use the NAERG Guide 161, 163, 164, 165, or 166. These recommend an initial evacuation distance of 100 meters for a large spill and 300 meters if a fire is involved. If explosives might be involved, move at least 1,500 feet away.

In transportation events or when a device is observed, a clue that radioactive material might be involved is the presence of heavy lead or other shielding such as concrete, plastic, water, or other metals. These are used to reduce the radioactive emissions to protect the persons transporting the device, and more importantly, make locating the device more difficult. If such is observed, an immediate evacuation and retreat by public safety personnel is needed. In a rare case, the magenta or black radiation symbol (propeller shape) on a yellow or white background indicates radioactive material is present. Outside assistance in the form of the bomb squad, federal assets, state radiological health, and health physics personnel are needed. If feasible and health physics personnel are present, they should direct decontamination, and make recommendations on exposure level and times allowed in the danger area if an entry is to be made. Responders should wear at least Level C protective clothing with APR (see Figure 99).

Figure 99. Radioactive material transportation placard.

Anyone potentially contaminated should doff their clothing and wash with soap and water. No one should drink, eat, or smoke in the danger area. When feasible, water runoff should be controlled. Keep those contaminated segregated until decontaminations can occur. Expedient decontamination can be accomplished by removing clothing should emergency medical treatment for life-threatening injuries be needed.

Radiation Exposure Devices

Another potential method to use the exaggerated terror from radiation-related events is an exposure device. This is where a radioactive source is placed in a setting in which a person or group of persons is exposed and they accumulate external doses of radiation. This usually targets one person or a small number of persons. One case involved a radiographic source planted in the target's work chair. The device was detected shortly after it was placed, as someone noticed the source was missing and began a search. One scenario that would create terror would be to locate a source in a heavily populated site, such as a stadium, during an event. Following the event, the information about the device and its placement is released. Officials would find the device and confirm its presence. The probability of a large number of persons receiving a substantial dose of radiation is small, yet the ensuing confusion and fear would result in most of those attending the event to seek medical assessment.

Chapter 9

BOMBERS AND CASE STUDIES

Although no universal profile of bombers exists, some commonalities can be seen in terms of behavior and motivation for bombings. Many classify bombings according to perceived motivation. These range from juvenile experimentation to terrorist suicide bombers. One noteworthy distinction of this classification is that many of the bombers may fall into several categories. In the US, the typical bomb will be a pipe bomb using smokeless or flash powder with a burning delay. The typical incendiary device, also classified as a bomb, will be a glass container of a flammable liquid, usually gasoline, with burning wick.

What has occurred is the responding officer can no longer expect these as typical devices. A sophistication of devices and improved explosive power has occurred. The Internet has provided a plethora of information allowing even juvenile experimenters to build sophisticated devices.

One motivation is the experimenter. This is usually a young juvenile male, but some immature adults may fall into this category. Their devices start as crude devices, usually an incendiary or crude pipe bomb with both using burning delays. The author has encountered juveniles as young as six years old experimenting, building, deploying, and exploding pipe bombs and incendiary devices while a 12-year-old manufactured TATP (see Figure 100).

Figure 100. A juvenile manufactured this Molotov cocktail. The juvenile managed to burn down a shed at his home.

Another common device is an overpressure device using HTH and isopropyl alcohol, or any corrosive material and metal, in a frangible glass container or plastic bottle. Known as a bottle bomb, these devices do present a hazard at close range. These explosions usually occur in a vacant lot, wooded area, or other open area (see Figure 101).

However, these devices are crude; they are functional and can cause damage and injury. The older the experimenter, the more sophisticated the device will be. Older juveniles may produce very sophisticated triggering devices using radio-controlled devices with a standard explosive such as smokeless gunpowder or homemade flash powder. What the Internet has prompted is that older juveniles and young adults are manufacturing sophisticated triggering devices coupled with the manufacture of primary high explosives such as TATP or HMTD. The use of legally purchased commercial explosives such as Tannerite or Star Targets has become popular. These can be adapted as a filler for destructive devices easily.

In many instances, the thrill of triggering the device may become tame. The experimenter may progress to a vandalism motivation where the target, usually public property such as a school, mailbox, or public building is targeted. Some authorities suggest that most crimi-

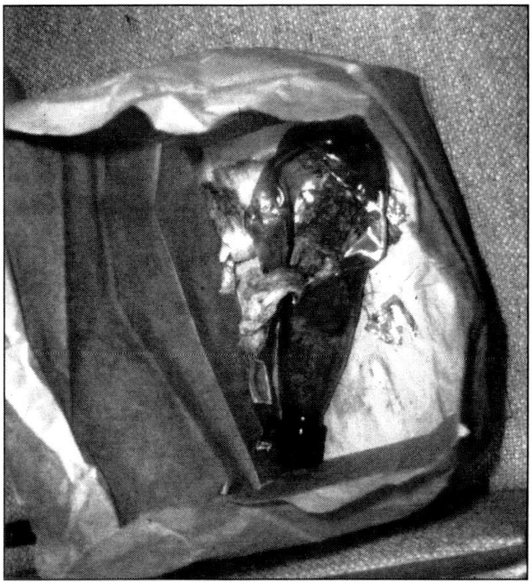

Figure 101. This bottle bomb was recovered and three young adults arrested and convicted for criminal possession of explosives. They had detonated more than a dozen of the devices.

nal mischief events involving explosions where property damage is small are committed by juveniles or young adults. Another variety of progression is the bomber may feel ostracized and target individuals or an institution, such as a student or a school, for attack. This is revenge motivation. This will usually be a juvenile or young adult. However, this motivation can be attributed to older more sophisticated bombers such as Walter Leroy Moody, who targeted a federal judge along with the legal system, and George Metesky, the Mad Bomber, who targeted Consolidated Edison.

Professional or mercenary bombers produce efficient and functional devices to accomplish a specific mission. These individuals will likely gather intelligence and spend time studying the target. The same or similar type devices may be used. The devices will be placed to cause the maximum amount of damage or injuries. This may be a series of bombings. The use of radio-controlled devices is not uncommon.

These individuals will likely use stolen commercial or military explosives with a very reliable triggering system. Some authorities classify terrorists within this category while others reserve a category as an ideological bomber. An example of an ideological bomber is Eric Robert Rudolph. He attacked Centennial Park during the Olympics in an attempt to embarrass the US government. Rudolph's grievance with the government was the perceived support by government of abortion and those living alternative lifestyles. He continued with other attacks. Terrorists from Middle Eastern countries could be classified as ideological in nature. Their use of radio-controlled devices has become widespread. Some of the designs used were acquired from designs used by the IRA.

The radicalized homegrown violent extremists, whether religiously driven or motivated by other hate, can be considered ideological bombers. This category is growing rapidly as US citizens go to locales where conflict based around their ideology is in progress. There these individuals train, engage in combat, and come back to the US with their beliefs strengthened. These individuals usually operate alone but may operate in pairs. They can target any facet of the US society with bombings. An example that some suggest meets the criteria is that of the Boston Marathon bombers.

Another category of bomber is the psychopathic. This individual has little visible external reasons for his or her attack. This individual is likely mixed with the ideological bomber, but the motivation may remain obscure. An example is the UNABOMBER who used bombs to attempt to change society from its addiction to technology to a simpler lifestyle. This was not apparent until his manifesto was published. Theodore Kaczynski hand built sophisticated devices to support his cause. Bombers may fit into several categories.

EXPERIMENTATION–This is usually juvenile offenders who are immature and curious. Their motivation may be from media portrayals or from information from the Internet. The construction and detonation of a device provides a sense of achievement, "bragging rights" to others, and excitement. These devices may be crude, but the author has seen 12-year-olds build substantial pipe bombs and manufacture TATP. Most devices are crude using smokeless gunpowder from firearm cartridges or flash powder from fireworks. Since incendiary devices are easier to manufacture, Molotov cocktails are common. Most devices will have a burning delay. Targets are usually mailboxes or abandoned structures. Bottle bombs are common with juveniles because of their ease of manufacture. The use of corrosive materials and aluminum foil or HTH and isopropyl alcohol are common ingredients. The three-liter bottle bombs, when constructed correctly, make an impressive explosion.

OLDER EXPERIMENTERS–these individuals may manufacture gunpowder, TATP, or HMTD. Part of the process is proving the individual can make the explosive. Sometimes this group may steal or purchase smokeless powder. They will progress from experimentation to vandalism as the explosion will bore them and the excitement phase will challenge them to become more bold and destructive. The common device is a pipe bomb and some will use a sophisticated initiation system. These devices will be progressive in size and sophistication. Some will use sophisticated initiation systems such as radio-controlled devices. Those over age 21 will likely purchase commercially available binary explosives such as those made by Tannerite and Star Targets. YouTube is a valuable source of information and possibly evidence. Several cases have been solved by watching the video of the event posted by the bomber on YouTube.

VANDALISM–this group is usually populated by the experimenters who have become bored with the detonation of devices in vacant lots or woods. They then target mailboxes, abandoned structures, vehicles, and perhaps a school building. The excitement comes from the destruction of property. The motivation is the thrill and excitement. However, some may progress to the revenge motivation.

REVENGE–this motivation has a large contingent of younger individuals, but older individuals may be part of this group. They seek retribution for a perceived wrong. The wrong may even be imagined. The target is usually an institution but may be an individual. The age of the offender is usually tied to the power and sophistication of the devices. As a rule, the more sophisticated and powerful the device, one can suspect an older offender. The attack on Columbine High School by Eric Harris and Dylan Klebold can be characterized in part by revenge. The Mad Bomber, George Metesky, can be classified as a revenge attack against Consolidated Edison for its firing of him. The author worked a serial bombing case in which two dissatisfied students targeted the school with several small bombs detonated during class to disrupt classes. The two females, unusual for bombers, were dissatisfied with grades they had received. They targeted trashcans, flower planters, and other physical objects such that the likelihood of injury was small but the noise and physical damage disrupted classes and served to terrorize students and staff. They were caught and prosecuted.

FOR PROFIT–these bombers usually perform the task as a sideline. It may be a component of organized crime or be performed at the behest of a person seeking revenge. Some are used to assassinate individuals, competitors in the organized crime setting, or to collect insurance from a destroyed structure. On some occasions, the bomber will be employed to eliminate a spouse or business partner. The key is the bomber is motivated in part by the remuneration for the act. On rare occasions, a first-time bomber attempting to collect insurance may populate this group with a crude device. Incendiary devices to destroy a structure are a common finding in the fraudulent collection of insurance.

RECOGNITION–this is the bored volunteer firefighter or police officer who uses a bomb or incendiary device to create excitement. They are the individual who discovers the device or arrives first on the scene. They want their actions to be perceived as heroic. The FBI in the Centennial Park Bombings believed security guard Richard Jewel to be a suspect in the bombing and focused on him while the suspect, Eric Robert Rudolph, planned his next bombing. This behavior is more common among volunteer firefighters than other public safety groups.

LONE WOLVES–this is another category, which has overlap into several fields of motivation. This could be revenge as in George Metesky, psychopathic behavior mixed with revenge in the case of the UNABOMBER or Eric Robert Rudolph, Timothy McVeigh, or the shooter in the Aurora, Colorado Theater shooting James Holmes. Holmes left multiple high-explosive booby traps in his home for police investigators to "trip." These individuals leave little signature during the planning phases and are difficult to detect. This group commonly uses covert delivery methods and may vary device construction as Eric Robert Rudolf did. This group will likely use high explosives, either stolen from a construction site or homemade TATP or HMTD. This group tends to produce more serial bombers. Their reasoning may not be apparent as in the case of the UNABOMBER (until his manifesto was published).

Individual Bombers

George Metesky, the "Mad Bomber" of New York City, is alleged to have constructed at least 30 bombs, deploying and detonating 20. His motivation was his firing from Consolidated Edison over a workplace injury. Metesky varied his targets with theaters, telephone booths, libraries, and train stations as a few of his targets. His 16-year bombing campaign kept residents of New York City on edge for almost 16 years. His campaign was well planned and executed. Metesky left little detectable signature but like UNABOMBER, letters Metesky wrote to Consolidated Edison were matched and led to his capture.

Ted Kaczynski, the UNABOMBER, conducted an 18-year bombing spree with 16 known bombs. UNABOMBER lived in a primitive cabin, hand-built his bombs, used the US Mail, or traveled by bus to hand deliver his devices. The hand building of his bomb components, avoiding retail purchase, and using cash when such was required, kept his signature very small. His delivery methods were anonymous and difficult to trace. Kaczynski opposed technology and the resultant effects of such on society. UNABOMBER demanded his antitechnology manifesto be published and the reading of the manifesto by a relative led to his identification. The amazing part of the Mad Bomber and UNABOMBER's campaigns is very little money was spent and large law enforcement resources could not locate or stop the bombings. Almost the entire law enforcement infrastructure of the United States targeted UNABOMBER but failed to detect and identify him.

This same low-profile, lone wolf style allowed Timothy McVeigh and a few confederates to construct and deploy the Oklahoma City bomb. The key is a low profile with only one or very small number of persons working to deliver an effective and low-cost bombing campaign. Another concept used by terrorist groups is "Leaderless Resistance." This is a central "leader" using the media or Internet to encourage supporters to take individual action. The instructions are produced carefully to allow deniability of the leader regarding any attacks, which occur from the rhetoric. This concept reduces signature as most terrorist organizations have a large support base, logistic operations, and even a recognizable leadership cadre. Leaderless resistance avoids these signatures.

Targets

In many instances, the target of the bombing may be significant. It may reveal the motivation of the bomber or group. In many instances, the attraction of the media will be a vital factor as it magnifies the terror produced by the bombing. This concept requires that important symbolic targets be attacked. If the attack can produce multiple targets, so much the better, as this attracts media attention. The Centennial Park bombing and the Boston Marathon bombings are examples of a symbolic target, which produced a large number of casualties (although the number of deaths was small). The media attention was intense and extensive. This literally paralyzed part of the northeastern US following the Boston Marathon bombing.

A series-bombing tactic, as used by the Mad Bomber and UNABOMBER produced a lower intensity attention but in doing so kept the public's attention for an extended period. Using anonymous methods such as covert delivery or delivery by US Mail or package carrier further conceals the identity of the bomber. The IRA and Basque ETA used series bombings very effectively in Northern Ireland and Spain to disrupt public events, cause fear, and create economic damage while attracting extensive media coverage. Eric Robert Rudolph's bombing campaign evoked fear in those using and working within abortion clinics.

Hard targets, those well protected, are likely to be avoided and soft targets attacked. The shift from attacks on better-protected government structures such as embassies, courthouses, and other symbolic structures has shifted to the softer targets. The Earth Liberation Front has used this tactic to attack building complexes under construction and sport utility dealerships with incendiary devices. In some instances, the device construction and target may lead to a signature, which identifies a single bomber or single group. Other soft targets include theaters, sporting events, transit systems, nightclubs, schools, hospitals, and hotels.

An example of a bombing attacking a soft target is the Bath Michigan School bombing of 1927. Andrew Kehoe, upset with the school tax and perhaps psychopathic, detonated several hundred pounds of dynamite in the Bath School. This followed killing his wife and destroying his farm. He drove to the school following the bombing in a vehicle filled with a large quantity of dynamite and detonated it, killing additional persons including the school superintendent and at least one student who had survived the initial bombing. The initial explosion killed at least 36 students and two teachers.

The Sterling Hall Bombing of 1972 on the University of Wisconsin–Madison campus involved a large vehicle borne improvised explosive device (LVBIED). This explosion only killed one person and injured others. Some 2,000 pounds of ANFO explosive are alleged to have been used by an anti-Vietnam war group. The buildings housed research facilities for the US military. This same group also stole a private plane and attempted to drop improvised aerial bombs on a munition factory.

In some circumstances, the attacks may be used to distract or acquire assets. Bombing have been used to distract law enforcement while a bank or armored car robbery occurs. One aborted bombing in Norco, California in early 1980s led to one of the more intense police shootouts in US history. The robbery group's bomb failed to detonate on a gasoline pipeline, but the group proceeded with their robbery. In the ensuing shootout, more than 1,000 rounds were fired killing one deputy sheriff, injuring more than a dozen officers, shooting down a police helicopter, and destroying more than 30 police cars. The suspects used shotgun-launched grenades during the pursuit.

In 2005, another bombing, which has an unclear motive, a student of the University of Oklahoma, detonated a person-borne improvised explosive device (PBIED) outside the Oklahoma Memorial stadium killing himself. Speculation existed that this was not a suicide as the explosive used is alleged to have been TATP, which is difficult to manufacture, but was an attempt to conduct a suicide bombing within the stadium. Controversy still exists regarding the motive.

In recent years, an attempted PBIED deployment occurred in a major metropolitan area, but law enforcement shot the suspect. The suspect was not known to be carrying an explosive device until EMS began emergency care and discovered the device. Two events in schools have had student bring PBIEDs within the school. One device was not functional and the student was restrained by school officials while in a second case, the suspect was able to deploy a hand-carried IED but was detained and restrained by students and a teacher prior to detonating the PBIED.

Terrorist groups prefer bombing attacks as they can be conducted without personnel present, are effective, and not expensive. The purpose of the bombing will likely be to produce the largest number of casualties feasible to attract media attention and demoralize the public. The signature to conduct such is small and arson and bombings produce little traceable physical evidence if precautions are taken.

Warnings of bombings are not common. However, in the current environment, one cannot ignore bomb threats. Bomb threats can be used to fatigue responders with multiple false alarms followed by a bombing attack. Even bomb threats are costly as employee time is lost and the bomb threats disrupt the function of the structure or facility targeted.

Serial Bomb Threats: Offender Profiles

The author while a bomb squad commander conducted data collection regarding those involved in making threats over a six-year period. The area served was roughly 5,000 square miles with a population of more than 300,000. The bomb squad responded to more than 550 calls during this period and made more than 60 bomb and bomb threat-related arrests. An interesting aspect is no bomb threat was associated with an IED.

A database is maintained on the type threats and persons arrested. From this database a general offender profile was developed from the 185 threats reviewed and arrests made. The known offenders, 57, were examined and eight had made multiple bomb threats.

An interesting trend was noted in that most bomb threats in this were revenge threats with thrill seeking the second choice. The thrill seekers tended to younger whereas, the older offenders made more threats and were motivated by revenge. Domestic disputes involving employees or labor problems with management played a role in several of the events. Most threats were made from nonactivated cellular telephones, which can still dial 9-1-1 but will not display a valid telephone number. This makes tracing the calls very difficult. Law enforcement and government facilities were common targets followed by large industry and schools. Bomb threats and bombings are normally thought of as a crime committed by males, usually white and young; however, females composed 38% of the serial offenders, and blacks composed 25% of the offenders. Five of the eight offenders had mental health histories and every offender had prior arrests.

Table 9. Overview of the Threats Made By the Known Serial Offenders

Sex	*Race*	*Age*	*Method*	*Number and Target*	*Motivation*
Male	White	17	Cellular	Five Commercial	Thrill+
Male	White	35	In Person	Seven Government targets and a Nuclear Facility	Revenge*
Female	Black	38	Telephone	10 Law Enforcement	Revenge#
Female	White	45	Writing	Four Law Enforcement	Unknown=
Male	Black	48	Telephone	Eight Government	Revenge=
Male	White	25	Telephone	10 Restaurant	Revenge
Male	White	23	Cellular	72 Poultry Processing Plant	Domestic
Female	White	21	Cellular	15 College	Thrill
Totals by Sex	Totals by Race	Totals by Age	Totals by Method	Totals by Target	Totals by Motivation
5 male 3 female	6 white 2 black	Age range 17 – 48 Average Age 32	Cellular 3 Telephone 3 In Person 1 In Writing 1	123 targets 14 Police 99 Commercial. 10 Govt.	20 Thrill 35 Revenge 72 Domestic 4 Unknown

+Suspect is developmentally delayed mentally.
*Suspect has an involuntary mental health commitment, threats taken seriously as owner of numerous firearms and former firefighter.
#Suspect has been involuntarily committed for mental health issues and has a conviction for arson.
=Suspect has a history of mental health commitments.

Table 10. Summary for Juveniles Nonserial Known Offenders Making Bomb Threats

Sex	*Race*	*Age*	*Method*	*Number and Target*	*Motivation*
9 Male 4 Female	9 White 4 Black	Range 10 – 7 Average Age 15	6 In Person 3 Cellular 4 Written	School 12 Residence 1	Thrill 9 Revenge 4

An interesting observation was the number of females involved and the large number of threats made in person. As expected, most threats were for thrills and a few were for revenge for a perceived problem with a school. As expected for juveniles, most threat involved schools. Several hoax devices were found coupled with these threats.

Table 11. Summary for Adults Non-Serial Known Offenders Making Bomb Threats

Sex	*Race*	*Age*	*Method*	*Target*	*Motivation*
19 Males 6 Females	10 White 15 Black	Age Range 18 – 61 years Average age 36 years	12 In Person 13 Telephone	3 Police 3 Homes 2 Business 4 Health 13 Government	19 Revenge 6 Unknown

Adult bomb threats followed a more common pattern with most offenders white males. The age range was surprising with a person as old as 61 making a threat with the average age range for offenders in their mid-thirties. Again, the common theme for motivation was revenge with domestic disputes playing a role in the threats. What was surprising was the number of threats made in person. Almost half the threats were made by the offender and were made in person. Several offenders have been arrested by law enforcement for the construction of bombs and incendiary devices. Their profiles are listed below. Simple bombs consisted of incendiaries or small nonmetallic pipe bombs using gunpowder. Most incendiaries were either bottle bombs using a caustic material and metal foil producing hydrogen gas and exploding, or the traditional Molotov cocktails using a flammable liquid in a glass container with textile wick.

Table 12. Incendiary Devices

Sex-Race	*Device*	*Age*	*Delivery*	*Target*	*Motivation*
Male (2) White	Molotov	8	Hand	Residence	Experiment
Male (1) White	Molotov	17	Hand	Residence	Thrill
Male (7) White	Bottle Bomb	18	Hand	Residence	Revenge
Male (1) Black	Bottle Bomb	20	Hand	Commercial	Thrill
Male (1) Black	Molotov	26	Hand	Residence	Revenge
Male (1) Black	Molotov	27	Hand	Residence	Domestic
Male (1) Black	Molotov	37	Hand	Residence	Domestic
Male (1) Black	Molotov	43	Hand	Residence	Revenge
Male (1) Black	Molotov	44	Hand	Residence	Revenge
Male (2) Black	Molotov	46	Hand	Residence	Revenge

The simple incendiary makers were male with an age range of eight years to 46 years with the average age 22 years. Every suspect delivered the devices by hand and revenge was the more common motive. The younger offenders did, however, experiment making the devices. The racial makeup was 56% white and the remainder black. None was noted to have prior mental health commitments. Offenders arrested for construction and use of pipe bombs or incendiary were profiled. Most devices were simple with either a fuse ignition system or electrical ignition system using smokeless or black powder in a metallic or plastic pipe. One device used a mercury switch, battery, and blasting cap with gunpowder. This device produced a fatality. The offenders were males, with roughly 80% white with an average age of 24 years. The number of injuries produced was one fatal injury and nine less serious injuries requiring hospital treatment or hospitalization with surgery.

Table 13. Pipe Bombs

Sex-Race	*Age*	*Delivery*	*Target*	*Motivation*	*Injuries*
Male (1) White	17	Hand	Residence	Revenge	None
Male (1) White	28	Hand	Residence	Experiment	None
Male (3) White	17	Hand	Commercial	Experiment	Three, two minor requiring hospital treatment, third serious, hospitalized with surgery
Male (1) White	30	Hand	Clandestine Drug Lab	Booby-trap	None
Male (1) Black	35	Package	Residence	Domestic	Fatal
Male (1) Black	23	Hand	Unknown	Unknown	None
Male (2) White	18	Hand	Commercial	Vandalism	None
Male (1) Black	28	Hand	Vehicle	Revenge	Serious, hospitalized with surgery
Male (1) White	36	Hand	Unknown	Unknown	None
Male (5 White	26-48	Detonated) during transport	Unknown	Unknown	Moderate

Suicide Bombers

Perhaps better described as a personal-borne improvised explosive device (PBIED) rather than a suicide bomber, these devices are elegant in that an escape plan need not be constructed. Some authorities have described these as a "poor man's cruise missile" as the target can be specific or used an area weapon as needed. These devices can be hand carried, worn, inserted in a body orifice, or even surgically implanted. The maximum weight of a device varies from about 25 to 50 pounds depending upon how the device is carried or concealed. These devices usually contain a high explosive such as PETN or RDX but may contain a homemade explosive such as TATP or HMTD. On occasion, low explosives may be used.

Enhanced fragmentation in the form a pipe bomb, metal fragments, nails, ball bearings, or similar items may be used to enhance lethality of the device. Some efforts have been made to reduce the metal signature of the device by using PVC or plastic pipe, minimal wiring, and glass marbles as fragments to defeat metal detectors. In suicide bombs, fragmentation will be the primary agent of injury.

Placing the bomb on a person can allow the individual to circumvent or breach security measures to gain entry to a sensitive area. One mode of attack is to use the lines of individuals attempting to pass through security as the target. In many instance, this densely populated area has little or no protection. Follow-up attackers wearing PBIEDs can breach security during the confusion following the first detonation. The lethality of the suicide bomber can be improved by arming the bomber with hand grenades and firearms. The PBIED becomes the weapon used last once a sensitive area is penetrated (see Figures 102 & 103).

Another tactic is to detonate a device, have additional suicide bombers hide among the injured individuals and await first responders. Coupled with this tactic might be additional persons wearing PBIEDs to attack the first responders while others attack the medical receiving facilities during the ensuing confusion. Using multiple attackers at a variety of locales attacking soft targets has been successful in the Mumbai-style attacks. The simultaneous attacks drain resources and provide fewer security personnel that the bombers must defeat. The attackers in the Mumbai event were well trained, and heavily armed with firearms and hand grenades. Responders should always

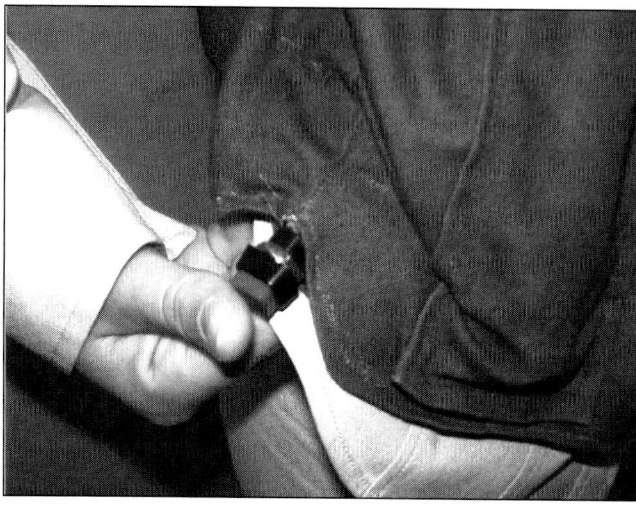

Figure 102. A toggle switch of the type used to detonate a vest type PBIED.

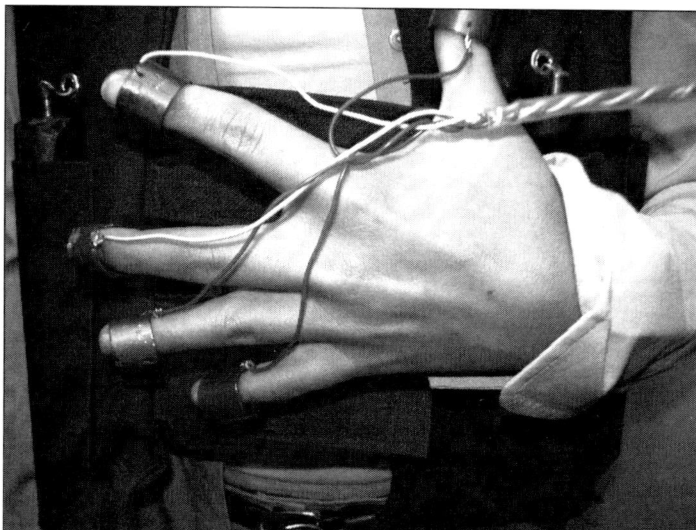

Figure 103. If the fingers come into contact with each other, the metal bands complete the firing circuit.

consider that any explosion might be a ruse to target responders or the beginning of a swarm attack (see Figures 104–106).

No reliable profile exists for suicide bombers, but many are young males with age ranges up to 50. Females have been used and pregnancy is a good simulation, which allows more explosives to be car-

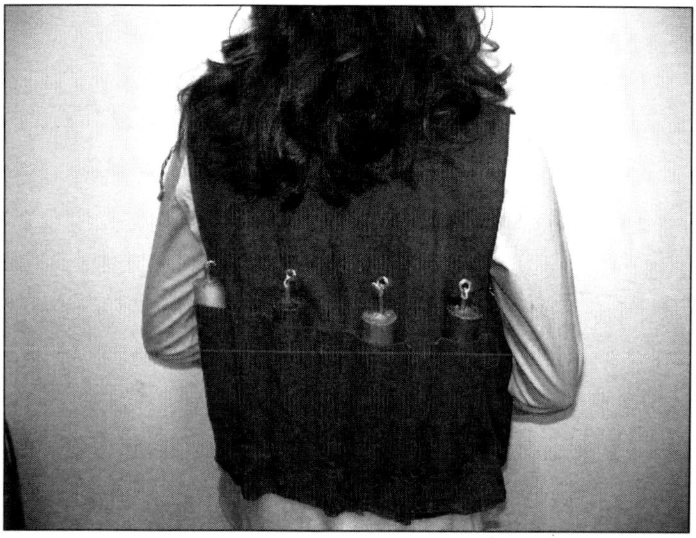

Figure 104. A rear view of the vest. Note the multiple sticks of high explosive dynamite.

Figure 105. The vest has a antiremoval device (paper clip holding a circuit open) such that removing the vest will complete the firing circuit.

Figure 106. Note the multiple sticks of dynamite.

ried and can account for the awkward movements they create. Official uniforms and VBIEDs using stolen or cloned emergency vehicles have been used. Usually clothing will be used to conceal the device. As seen with the "underwear bomber," the bomb may be concealed in undergarments such as a woman's bra. A new tactic is to conceal the explosive in the rectum or vagina. These orifices have long been used to smuggle items and the logic of using them is clear. However, this does limit the amount of explosives that can be carried. The body does absorb some of the explosive power of the blast but bones do make good fragments.

Further, the blood and tissue increase the biohazard risks. An unusual option is to take the body orifice concealment further and use surgical implantation. The size device becomes much smaller, but this might defeat some security measures. Most security agencies will be reluctant to and may not be equipped to x-ray the person. If liquid explosives are used, they may not even x-ray as liquid explosives as some are postulated to have a density similar to tissue. Medical devices such as a pacemaker or insulin pump are used to house a detonator system or explosives. These items would appear on x-rays as legitimate. It becomes difficult to differentiate legitimate medical devices such as an indwelling IV catheter or IV port accessing the central circulation from an IED using liquid explosives. A person with fresh surgery attempting to enter a sensitive area should have the medical procedure confirmed. Other factors used to conceal a device are clothing that may be inappropriate for the temperature or climate, or may appear ill-fitting or baggy. Bandages and crutches have been used to conceal the explosives. Bicycles are another method in which the hollow tubes of the bicycle are filled with explosives and the suicide bomber rides to the target. Other options include hand-carried devices concealed in a brief case, package, and the bomber dressed as an official or delivery agent.

The facial expressions range from nervous to unresponsive with a blank stare. The bombers will likely have stiff movements because of the explosives carried and may make "confidence touches" similar to those carrying firearms. Those with explosives in the rectum or vagina may have an unusual gait. Other signs include bulges in unusual locations, visible wires, or batteries, or a handheld switch. Additional explosives may be concealed in a backpack, hand-carried items, or briefcase.

Suicide bombers may be the prelude to an attack or may follow an attack targeting responders. In rare cases, multiple bombers may target a single area. If one suicide bombing occurs, be alert for follow-up attacks. A method used to defeat security precautions is for the suicide bomber to enter the security area and jump across or run through security barriers. If confronted in the security screening area, the suicide bomber may detonate the device killing a number of persons in line and the law enforcement personnel present. This would then open the facility for additional attackers.

Standard security screenings will detect most devices, especially if coupled with hand searches of packages or suspicious persons. When feasible, an overwatch or high-ground observer with a long gun might facilitate stopping a suicide bomber if the rules of engagement are liberal. Layered security is essential to defeat a suicide bombing attack when such destroys the perimeter entry point. The rules of engagement must be clear and comply with legal parameters of the jurisdiction. However, such must facilitate engagement with deadly force of a person reasonably believed to be a suicide bomber. The policy must be coupled with training on the procedure and training both in the classroom and practical shooting training to assure a functional plan.

Consideration must be afforded to the probability of a detonation if a bullet strikes a bullet-sensitive explosive such as commercial dynamite or TATP. The explosive will likely be worn in the torso area, which is the traditional center of mass shot for law enforcement. Procedures may have to be flexible to allow head or pelvis shots. The priority remains to incapacitate a suspected suicide bomber before he or she can detonate the device.

If what is believed to be a suicide bomber or VBIED is detected, several issues should be considered:

- Containment and incapacitation of the suspected suicide bomber or vehicle
- Engagement with deadly force if such is within rules of engagement with a head or pelvis target
- Immediate evacuation
- Consider this as a prelude to additional attacks

In rare cases, negotiation may have to be undertaken with a suspected suicide bomber. If so, such should be:

- Conducted from at least 100 feet away under hard cover and not near glass
- If possible, the suspected suicide bomber should be covered by a tactical shooter with a long gun
- Expect an explosion and reaction time will not allow one to gain cover
- This may be a ruse to distract or concentrate security forces for another attack
- Do not close with a suicide bomber even if the suspected bomber is down as their explosive device will likely be booby trapped, on a timer, or may be radio controlled by a "minder" present in the crowd or nearby

If a suicide bomber attempts to surrender, consider:

- Such is a ruse to distract or concentrate security forces
- Have the suicide bomber open his or her hands and if wires or a switch are visible, this likely is suicide bomber

- If compliant, have the suicide bomber to remove clothing and turn 360 degrees, look for wires or devices visible in the rectal or vaginal area, look for fresh scars, tubing or wires protruding from the body
- If what appears to be a suicide bomb is still present on or within the body of the suicide bomber, have him or her lie down and spread his or her arms and legs, do not approach the suspected suicide bomber but call for bomb squad support
- If the suicide bomber appears to have discarded the visible suicide bomb, remember it may still be present concealed in the rectum, vagina, or surgically implanted
- Wait for bomb squad support and do not allow the suspected suicide bomber to approach officers and officers should not approach the suspected suicide bomber

If an explosion has occurred and it appears to have been a suicide bombing:

- Other suicide bombers or another attack may be imminent
- Expect parallel or swarm attacks similar to the Mumbai attacks
- Suicide bombers may be among the injured or may be feigning injury to reach the hospital to detonate the device
- Do not concentrate security forces solely at the bombing site
- Increase security at other sites which may draw crowds such as hospitals

If the suicide bomber is not incapacitated, he or she will likely complete his or her mission. This problem has spread throughout the US and is an issue law enforcement and security forces must consider in every venue. The negative aspect is that the use of deadly force in a suspected suicide bomber not carrying a device will generate negative media coverage and public outcry. This was seen in an earlier shooting in the United Kingdom where officers, following protocol, engaged a suspected suicide bomber not carrying a device. Debate still exists as to whether or not this was a test of security forces or a "dry run bombing."

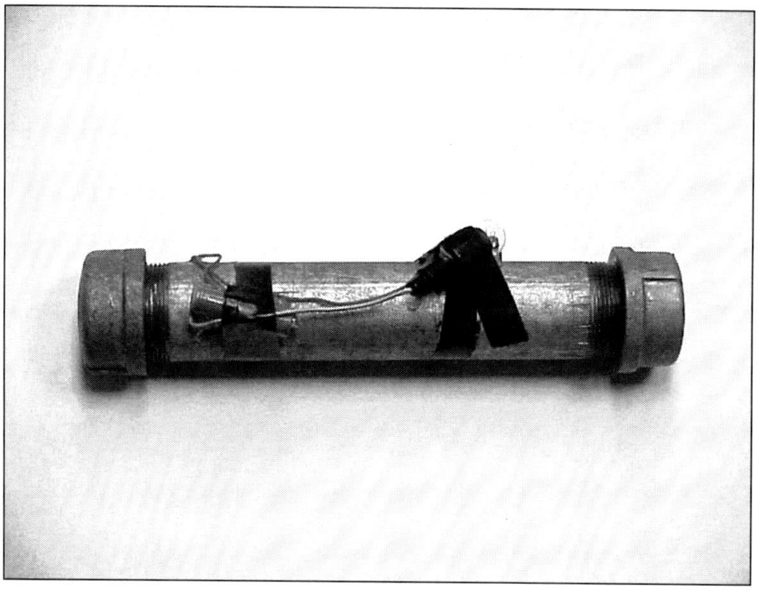

Figure 107. This device uses a mercury switch from a trunk lid light system. The blasting cap is contained within the pipe.

Domestic Bombings

These bombings are sufficiently different from other bombings to warrant special attention. Open source data from the BATFE indicate that these are characterized as revenge bombings by the male partner (female bombers are rare). White male bombers are more common, although the author was involved in an investigation with a fatal domestic bombing conducted by a black male.

In this case, the bomber purchased the metallic pipe and end caps from a local hardware store, smokeless gunpowder from a sporting goods store (he told the owner he was feeding the gunpowder to his dogs to make them mean), and mercury switch, batteries, and wiring from a local electrical supplier. He used a book from the local library on how build bombs. The bomb was handed to the target's daughter in a container. The curious 11-year-old opened the container, causing the device to detonate and killing her. The bomber was remembered at each purchase locale and the library because of his nervous behavior. The bomber was executed several years later (see Figures 107 & 108).

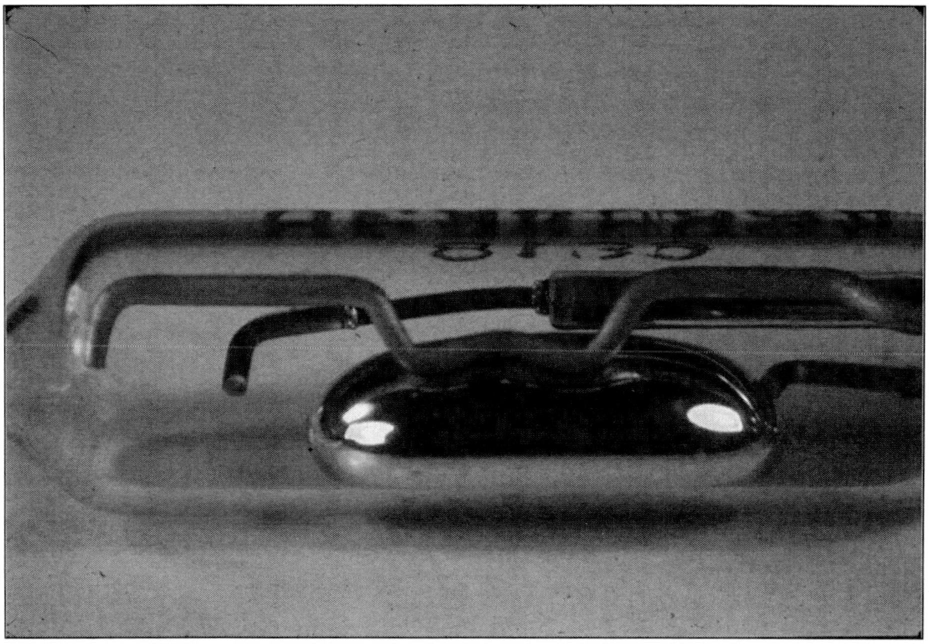

Figure 108. A close-up photograph of a mercury switch. These are becoming difficult to locate because of the hazardous materials risk of mercury.

These bombings are likely to occur at the residence or workplace. The bomber will have sufficient motivation to not be concerned with collateral casualties or damage. The sole goal will be to maim or kill the former domestic partner. In rare cases, those family members or friends, who have supported the target, may also be bombing targets.

This same data indicates that the fatality rate is very high with roughly half the targets fatally injured. This is a higher rate than typical bombings as the devices tend to be well built and are target activated. As with most bombings in the US, the device is a metallic pipe bomb 10 to 16 inches long containing smokeless gunpowder. The attacks are well thought out. Most devices are fragment enhanced and a few use high explosives. The devices will likely be hand delivered disguised as a mail or other delivered package. The device will likely use an electrical ignition system.

Some of the more common factors seen with domestic bombings include:

- Bitter dispute
- Female initiated divorce

- Serious child custody disputes
- Male is placed in a serious financial strain because of the divorce
- Property settlement issues
- Domestic violence has been involved and a protection order is in place

Other factors, which make these bombings significant, are the disregard for injuries and deaths of those involved. The bomber may not take efforts to conceal his identity with the only objective to kill the former domestic partner. The bomber knows the behavior of the target and schedule making the bombing easier. Those involved in a bitter divorce proceeding should be schooled to identify suspicious packages, not open any unexpected items, and use countersurveillance methods to detect surveillance.

Chapter 10

BOMB THREATS AND BOMB SEARCH PROCEDURES

The typical bomb threat is a simple, "There is a bomb in building." Is this a credible threat? The totality of circumstances must be considered. Is this a location with no controversy, no labor dispute, not a high value target, and not visible in the community, then likely this is not a credible threat. The call stating, "A pipe bomb filled with smokeless powder is concealed on the first floor. I used an old Baby Ben clock as a timer hooked to a battery with a Christmas light as an igniter. You have about 30 minutes to clear the first floor." This would be a credible threat, especially if the location is the subject of controversy, has a labor dispute, and is visible in the community.

How do the police, security, and facility leadership react? In the first scenario, several options are available:

- Take no action and hope no bomb is present
- Perform a search of employee areas and public areas
- Evacuate and risk exposing employers to external threats

The problem with taking no action is, if an IED is present and detonates, the leadership of the facility has substantial liability. This is the reason law enforcement should not make recommendations in bomb threat situations or state, "No bomb is present," when a search is negative. The key is that, even with a search, a well-concealed IED may not be discovered. If public buildings are involved or if a suspicious object is found, law enforcement may have to make a decision regarding evacuation and do the following:

- Gather intelligence
- Observe the site and determine if additional resources needed
- Determine what if any evacuations are needed
- Perform a search if needed
- Evacuate if needed
- Isolate any device found
- Stage other resources
- Take protective actions (evacuate, barricade, open windows)
- Observe the device
- Evaluate the device
- Make an assessment of options: Primary, secondary, tertiary
- Proceed with options
- Complete evolution
- Evaluate situation
- Deescalate site
- Obtain additional resources if required
- Debrief and critique
- Prepare report

One important facet of evacuation is to have a nearby evacuation area which is safe for employees. Some facilities have agreements with other nearby businesses to use their facility as a shelter. This is true in several circumstances regarding schools, which tend to be in proximity to one another. Another offsite evacuation and assembly site a distance away is needed. One school system uses school buses to evacuate students during an emergency including credible bomb threats. This occurs after school officials and law enforcement search the buses and offsite evacuation site some 10 miles away. However, this site has a large auditorium, is secure, and has substantial parking assets. Law enforcement from another city provides security and parents have been advised with a brochure on how to retrieve their children. The same school system uses nearby churches that students can walk to as a nearby evacuation site. This particular system has used this process on several occasions and found the plan worked well. This same plan has been used to evacuate when a suspected school shooter was near the school. Officers stopped the shooter two blocks from the facility.

Some facility leadership evacuates with every bomb threat. This is an absurd response. Blind evacuation may lead to a situation faced in Arkansas where school shooters positioned themselves to shoot students and teachers as they evacuated with a fire alarm one of the shooters pulled. Evacuation areas should be checked for threats such as prepositioned devices, VBIED, or shooters. Further, an adversary may be checking the response of the facility and law enforcement to the threat. Multiple threats fatigue the vigilance of employees and they may begin to ignore such.

An evacuation area with suspicious packages, unknown persons present, or a vehicle which does not belong, may contain a threat. If such is found, one should use a secondary evacuation location. However, in many cases, the employees or occupants of the facility have been evacuated prior to the arrival of law enforcement. If the bomb threat is telephoned, several steps should be taken:

- If caller ID is available, the calling number should be recorded
- If the call trace feature is present, it should be used
- What is said should be written on a bomb threat form
- If the call is recorded, it should be checked for additional information and
- The call should be made available to security and law enforcement
- If a party is named, where is the party

Bomb threat credibility assessment is easily performed and should be loaded to overreact. However, a systematic threat assessment will provide the facility leadership a guide in the decision-making process.

In the assessment below, a score of three or more suggests the threat is credible. However, the totality of circumstances must be considered.

LAW ENFORCEMENT AND SECURITY THREAT ASSESSMENT CRITERIA

Time and date call received: ______________ Caller ID or *57
Trace used? Yes No

Exact words used in the threat: ______________________________

____

Threat Assessment: (verbal and written threats)

Did the caller say or does the note read:

Where is the bomb?	Yes	No
When will it explode?	Yes	No
What does it look like?	Yes	No
What type bomb is it?	Yes	No
What can be done to stop it?	Yes	No
Was money demanded?	Yes	No
Was anything demanded?	Yes	No
Was trace or caller ID successful?	Yes	No
If successful is the party known?	Yes	No
Is the threat aimed at a specific person?	Yes	No

Score Yes 1 Point Each Total Score ______ *Score >3 Are Credible Threats*

Target Assessment Factors:

Prior bomb threat(s) in last year?	Yes	No
Employee-employer controversy?	Yes	No
A specific person named?	Yes	No
Specific person has domestic problems?	Yes	No
Labor dispute?	Yes	No
Government facility?	Yes	No
Government official named?	Yes	No
School?	Yes	No

Medical facility?	Yes	No
Recently terminated employee?	Yes	No
High-risk facility?	Yes	No

(Jail, police facility, courts, lawyer, judge, abortion clinic, hospital, physician, animals at facility, timber business, environmental related, media, international business).

Recent news story on facility or person?	Yes	No

Score Yes 1 Point Each Total Score ______ *Score >3 Are Credible Threats*

Usually the more information obtained regarding the device, the more credible the threat. A balancing test must be used to determine the perceived risk against the credibility of a threat and the likelihood of injury or death should an IED be present.

A confounding factor is if money or other demands are made. This threat may be designed to distract police or obtain money as in a robbery in which the robber threatens to use a device or leaves a package alleged to be a bomb.

If a successful phone trace is accomplished, the person the telephone is listed to makes another important factor in the decision-making process. Threats now come in many forms; they may come via email, mail, or in person. However, a systematic threat assessment will provide the facility leadership with a guide in the decision-making process. Federal Judge Vance in Birmingham was the target of a lethal package bomb sent by an unhappy defendant. Judge Vance had received threat letters regarding his decision. The bomb was sent with a return address, which led the judge to believe the package originated from another federal judge in Atlanta. The device was designed to detonate when opened. Even increased vigilance may be defeated as in this case.

Facilities such as government buildings, educational facilities, and public safety facilities are likely to receive credible threats. These facilities should consider using call trace features provided by the telephone company sometimes known as *57. However, with the advent of "throwaway" cellular telephones, this may only lead to an untraceable cellular telephone. Confirm the threat came in on an outside telephone line. In some cases, the threats are made locally via a closed telephone system. Some older internal telephone switches may not allow the extension in use to be detected. If the location of the telephone is discovered, it should be protected as a crime scene. Crime scene technicians may be able to lift fingerprints or in some rare cases obtain DNA. The area should be checked to determine if video surveillance is present and recorded. This could assist in identification of suspect.

Bomb threats in many areas are a felony and if the event is one in a series, then any information is important. If the same cellular telephone is in use, a search warrant or court order might be obtained to track the cellular telephone. Threats to use a weapon of mass destruction may fall under federal law. Requesting assistance from the Federal Bureau of Investigation should be a consideration.

Open source information indicates that few bombings are preceded by a threat. The more common purpose is to generate disruption, create a nuisance, attract media attention, and cost the entity time and resources. Law enforcement should attempt to conceal bomb threats from the media but must respond to media inquiries. The CEO of the law enforcement agency should establish a clear policy for the agency PIO to follow. Some agencies refer the inquiry to the facility threatened with the knowledge the facility may provide sensitive information to the media. Other agencies provide a "bare bones" response to the inquiry.

The traditional bomb threat checklist is useful as it can guide the answering person through the questions to ask. It may also assist in containing fear since it can focus the call recipient upon a task.

BOMB THREAT SHEET

Time the call was received: ______________________________

Where is the bomb located? ______________________________

When will the bomb explode______________________________

Is the bomb disguised? __________________________________

What can be done to stop the explosion? ___________________

Who is this? (This usually prompts a hang up) _______________

Write down exactly what was said. _______________________

First Responder

Questions for the first responder:

1. Was the call from an outside line? Was the Calling Number ID displayed? Has Call Trace (*57) been used? Is the line the call received recorded?
2. If the Call Trace feature was used, has the telephone company been contacted? Has the telephone been used since the bomb threat was made? Have you contacted the Telephone Company?

Questions to ask the facility leaders:

1. Has the facility received prior threats?
2. Is there any controversies concerning the facility?
3. Have any employees been terminated recently?
4. What does the leadership want to do in terms of a search or evacuation?

Bomb threats do not always come via telephone; they may be sent via email, via US Mail, or made in person. Those sent via email may be traceable if an IP address can be obtained. Written threats received by mail should be handled carefully as the envelope may contain an unknown powder and fingerprints and other evidence may be obtainable from the envelope and threat paper. In-person bomb threats present a different and more immediate threat. The person may have a PBIED or a firearm. Some agencies treat these threats as if the person has a PBIED until such can be determined.

If the suspect makes the threat and leaves, any items touched or handled should be isolated to be examined for fingerprints. If the suspect is determined not be carrying an IED, the on-scene officer should follow local protocol regarding questioning of the suspect. Many agencies treat this as an exigent circumstance and begin questioning the suspect by on-scene officers until investigators arrive. Officers should also determine that the threat made does in fact constitute a crime. One factor to consider is to isolate the person and evacuate away from him or her, or to remove the person from the facility.

The first responder will need to coordinate with facility leadership to obtain the needed information. If an evacuation has occurred, are the employees in a place of relative safety? Consider how susceptible the employees are to sniper attack or preplaced IED attack. Consider the threat may be distraction while another crime occurs. If feasible, the employees should be upwind and uphill from the facility.

The officer should determine if the employees performed a search of their work area prior to evacuation and were any other searches performed. Accountability of employees needs to be confirmed. Make certain no one is standing near or under glass. The first responder should follow local procedure but should consider the options available and what outside support might be needed. If a search is to be undertaken, then several searchers will be needed. If the threat is sufficiently credible, the deployment of a canine trained to detect explosives should be considered. In most locales, the bomb squad does not respond unless a suspicious object is found, but some may have electronic devices capable of detecting explosives. They may provide assistance using these devices. The officer should determine if the bomb squad is available and if needed, how long their response time would be.

Search Procedures

The blind evacuations following bomb threats to some facilities leaves the responding officer with a crowd of persons milling about while standing under glass with no visible leadership. Employees have probably left their personal belongings inside and did not perform a visual sweep for any out of place items in their work area prior to leaving. This presents several issues. If a search is needed, one finds personal items in a variety of locations, which any could be occupied by an employee backpack, purse, or briefcase. Officers will not know what belongs and what is suspicious. An employee will likely know his or her work area and would detect any suspicious item present.

Searches come in many forms. Limited searches may consist or employees visually sweeping their work area for items they did not bring in or do not recognize. These objects should be investigated by the leadership to determine ownership and if such cannot be determined, it should be treated as a suspicious object. Maintenance or environmental service employees should check external and public areas for unattended bags or items. Employees must be cautioned about items they do not recognize and instructed not to disturb them. The author has had employees bring a suspicious object to the command post rather than identifying the object, isolating the area, and reporting. Remember, a search may trigger the device and this factor must be weighed before determining to conduct a search placing the searchers at risk.

Public access areas should have priority, which includes the exterior of the building, especially near utility connections or sensitive areas. This should include within and around trashcans, plants, and flowerbeds. Inside the structure, hallways, public restrooms, lobbies, auditoriums, and waiting areas should be checked. Employees should look for any out-of-place item, which includes boxes, briefcases, purses, backpacks, or bags. Another source of bomb delivery is the mail or package delivery services. This is discussed later.

Room Searches

The presumption is the facility has been evacuated. Gather intelligence on any prior threats, threat groups, or labor issues if a commercial facility. If a detailed search is decided upon, then pair law enforcement personnel with facility employees when feasible. Maintenance and environmental personnel will usually have keys to most areas. The correct method to search a room is with two searchers and is time consuming. Such should be conducted in a systematic manner. Enter and familiarize oneself with the room and stop to determine the ambient noise level while standing back-to-back. Any out of the ordinary noises should be carefully investigated. The room is arbitrarily divided in half with one searcher taking each side. Desk level should be investigated first, followed by ground level, and finally the overhead. One must also search suspended ceiling overheads. Opening and closing drawers and cabinets should be performed carefully and slowly.

If a suspicious object is found, the search is stopped, the object isolated, and a report is made to the command post. Efforts should be made to identify the item as belonging to an employee. If such cannot be accomplished, the bomb squad will need to intervene. The problem one finds with schools is that they are usually full of backpacks.

If a search reveals no suspicious object, the searchers should mark the door (some agencies use wax pencils while others use a self-adhering colored paper dot). Searchers should thoroughly search every area. Some agencies use thermal imagers in hopes of locating an elec-

Figure 109. Opening doors may detonate the device. This device is motion sensitive with a mercury switch. This hoax device was found within a commercial structure.

tronic device heat signature. Be alert for out of place wires, fishing line, or string, which might be a booby trap. Make certain delivery points and sensitive areas are examined.

Vehicle Searches

Anyone with a minimal amount of training can be taught to locate hand placed bombs attached to vehicles. However, devices more sophisticated may be undetectable by a casual search. The initial effort should be to assure the surrounding area is safe. Survey the vehicle and its locale for any suspicious objects, wires, or other items such as bags, boxes, backpacks, or briefcases nearby. Some agencies use a thermal imaging device to look for anomalies. A walk-around inspection should occur without touching the vehicle. One must remember the device in a vehicle may be on time or is motion sensitive (see Figures 109 & 110).

Any nearby trashcans, dumpsters, boxes, planters or the like should be examined for an IED targeting responders. The vehicle exterior should be examined looking for wires, fishing line, strings, or any out of the ordinary. During the external examination, look for smears or

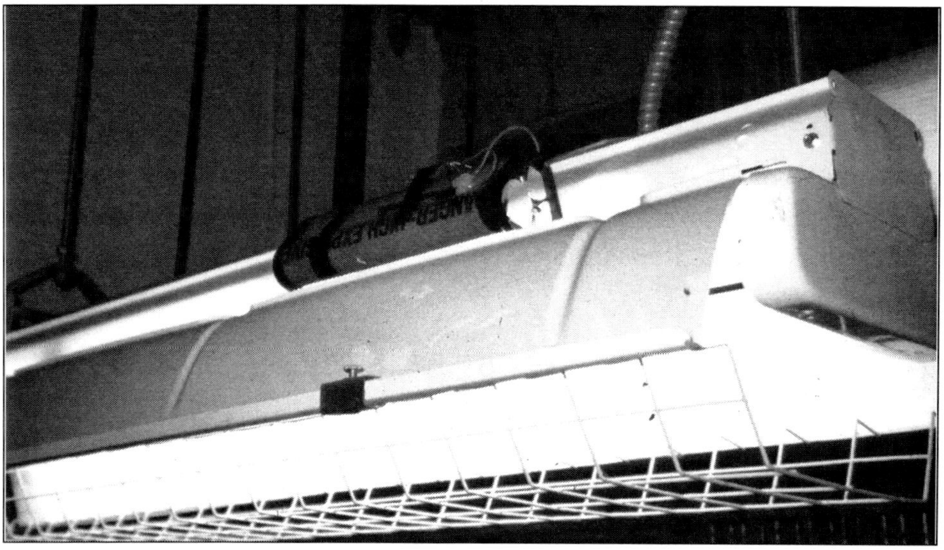

Figure 110. A hoax device with a motion sensitive mercury switch was well hidden within a commercial structure. The device was found several days after a bomb threat was made.

evidence an area has been touched. Examine the vehicle interior from the outside without touching the vehicle. Check the vehicle underside with a mirror or video system for any anomalies. Include the wheel wells and underside of the engine compartment in the inspection. Examine for smears, disturbed areas clean in comparison to other areas or loss of caked-on soil. Pay close attention to left side of the vehicle first before checking the underside of the fuel tank and catalytic converter. Remember the driver's side is a common target for bomb placement if the driver is the target.

If the search is negative, remotely open the right rear door. Examine the interior without entering using a remote mirror or video system. Open the left front door and remotely examine the interior paying attention to the driver's side. Repeat the process with the left side passenger door. If a hatchback or rear door exists, attempt to open it remotely. With some vehicles a remote can be used to lock or unlock doors, open or close some doors, and the gas flap, or start the vehicle. Use this remote from a position behind hard cover away from glass. Stay behind cover at least 60 seconds for each remote operation.

If feasible, open the hood remotely including the secondary latch. If feasible, examine the engine compartment with a mirror or video system without touching any of the fixtures. Pay attention to the driver's side and firewall. Look for extraneous wiring, areas of smears, and disturbed areas of dirt film from wires or components. Some bombers use wires attached to spark plugs or the car battery, as an electrical source to power the IED initiation system. Look for strings, fishing line, or wires attached to the IED.

If nothing is seen, open the hood fully and if needed, prop it open with a wooden or plastic rod. One should examine the seat surfaces for cuts or tears and palpate the surfaces for inserted objects. The underside of seats should be examined carefully. Remotely open the glove box or other interior compartments. If a trunk is present, make certain the trunk is carefully examined including any "wells" present and spare tire. Open the gas fill flap remotely and examine the gas fill area, remove the gas cap slowly feeling for resistance and looking for wires, string, or fishing line.

If reasonably certain no IED is present, operate the switches in the vehicle and move the vehicle through it gears. At any point a suspicious object is observed, or wires, string, fishing line are observed in locations they should not be present, or if any tampering is detected, isolate the vehicle and evacuate. Call for the bomb squad. Remember to stay behind hard cover away from glass.

Evacuations

If a decision is made to evacuate or if evacuation has already occurred, how far does one place the evacuees? First, the area the evacuees are in should be examined for safety and its distance. Small devices such as a pipe bomb do not require long distances if the device is within a structure. Some agencies adopt a rule of thumb that if a pipe bomb is present or the explosive weight is less than five pounds, they evacuate the floor on which the device is located and the floor above and below it. Horizontal evacuation should be at least 70 feet but preferably further away. Remember glass is deadly in the explosion setting so make certain that no one stands under glass. If the IED is located outside with no intervening structures or terrain features, the evacuation distance becomes further–roughly 1,200 feet for a pipe bomb. Larger devices require a longer distance (see Figure 111).

Accountability is important as is the employees examining their work area prior to departure and carrying personal belongings. The fa-

BOMB THREAT STAND-OFF CHART

Threat Description Improvised Explosive Device (IED)	Explosives Capacity[1] (TNT Equivalent)	Building Evacuation Distance[2]	Outdoor Evacuation Distance[3]
Pipe Bomb	5 LBS	70 FT	1200 FT
Suicide Bomber	20 LBS	110 FT	1700 FT
Briefcase/Suitcase	50 LBS	150 FT	1850 FT
Car	500 LBS	320 FT	1500 FT
SUV/Van	1,000 LBS	400 FT	2400 FT
Small Moving Van/ Delivery Truck	4,000 LBS	640 FT	3800 FT
Moving Van/ Water Truck	10,000 LBS	860 FT	5100 FT
Semi-Trailer	60,000 LBS	1570 FT	9300 FT

1. These capacities are based on the maximum weight of explosive material that could reasonably fit in a container of similar size.
2. Personnel in buildings are provided a high degree of protection from death or serious injury; however, glass breakage and building debris may still cause some injuries. Unstrengthened buildings can be expected to sustain damage that approximates five percent of their replacement cost.
3. If personnel cannot enter a building to seek shelter they must evacuate to the minimum distance recommended by Outdoor Evacuation Distance. These distance is governed by the greater hazard of fragmentation distance, glass breakage or threshold for ear drum rupture.

Figure 111. A card showing recommended evacuation distances for a bomb threat with a suspicious item or suspected item.

cility should have a recall method and method for communications at the offsite assembly areas. The use of cellular telephones and portable radios should be avoided, especially near suspicious items. However, if this is the only form of communication, it should be used sparingly.

Evacuate away from suspicious objects. Do not move suspicious objects. The author has had police officers and citizens bring suspicious objects, pipe bombs, and military ordinance to the police department. An event in Milwaukee, Wisconsin in 1917 killed nine police officers when a suspicious device was brought to the police department. The device was a bomb.

Invariably the media will become involved. The decision on what to release is important as the fewer the individuals aware of the threat, its content, and the reaction is important. This makes it more difficult for individuals to gain information in this regard and may avoid copycat threats.

Chapter 11

RESPONSE TO BOMBINGS

Bombings are easily accomplished, as they are inexpensive, technically simple, and anonymous. The materials are simple to acquire, and they garner media attention. The word "bomb" tends to evoke fear out of proportion to the threat presented. No escape plan is needed and the explosion itself can hamper evidence gathering.

The typical bombing in the US is a pipe bomb filled with smokeless gunpowder in a metallic pipe or a glass container with a flammable liquid ignited by a textile wick. Even so, these devices can be lethal as evidenced by the Centennial Park Bombing and UNABOMBER campaign.

The use of VBIEDs changes the scenario in the US. Although VBIEDs have been used in the past in the US, the Oklahoma City bombing riveted the public's attention. Many thought this event had to be the work of an international terrorist group. As the bombing emerged as the result of a lone wolf, the public's perception of bombings changed. This effect was further amplified with the 2001 aircraft attacks on the World Trade Center.

These bombings evoke fear and attract media attention as these events are sensationalized. The targets may be symbolic, but the shift has been toward targets with a dense occupancy to produce many causalities. The shift toward buses, trains, hospitals, night clubs, sporting events, daycare centers, and schools in lieu of well protected and government structures has made the job of the bombers easier. This change of targets means that literally any densely occupied location could become a target of any group with sufficient resources to construct and deliver a VBIED.

The media would have one believe that bombings are common and according to data in the public domain, about 4,000 bombings (which includes the use of incendiary devices) occurs in the US annually. The typical bombing is a small metal pipe bomb or Molotov cocktail attack. However, with the switch from symbolic government targets with "target hardening," one may see a shift to schools, hospitals, nightclubs, and theatres as evidenced by the attack on the Aurora, Colorado theatre with the suspect's apartment booby trapped.

Each attack, which is successful against a target with no or little protection, will cause the public and government officials to further extend resources. This makes information gathering and assembling intelligence crucial. The agency must be aware of the threat groups in its jurisdiction, their capabilities, and intent. This allows the deployment of resources in an efficient manner and may allow the disruption of a plot.

To be taken into consideration, according to open source materials, about half of the terrorist plots disrupted in the US came from information developed by local agencies through community policing programs. Several bombers and terrorists have been arrested by local agency officers from traffic stops unrelated to the mission of the terrorist. The index of suspicion should be high when any explosion is reported, as this may be the starting of a swarm attack or diversion for another crime.

The key to any response is to maintain personal safety. Toward this end, one needs to remember time, distance, and shielding. This concept emphasizes staying the threat are the minimal amount of time possible, staying as far away from any suspect items as possible (staying from near or under glass), and keeping any solid objects between the responder and suspect device or threat area.

- Consider the explosion may be a distraction or beginning of a large-scale attack.
- The explosion may be to attract responders into a scene to target them with firearms or secondary explosive devices.
- When approaching the scene, be observant for out-of-place vehicles, suspicious persons, and broken glass. Broken glass will indicate that overpressure or fragments affected the area.
- When feasible, survey the scene with binoculars looking for persons down, areas damaged, potential suspects, potential attackers.

- Be prepared for building collapse, loss of utility service, multiple casualties, persons fleeing the scene, and initial units are likely to be overwhelmed by the injured. The scene will be dusty and persons exiting will likely be covered in dust and debris. Broken glass will cause many injuries.
- Call for the bomb squad, mutual law enforcement aid, EMS. Urban collapse rescue teams may be needed if structures have collapsed.
- The blast area should only be entered only to save lives. Use a public address system to have the wounded who can walk come to a location for EMS to treat them. Have those who cannot walk shout or raise their hands. Only then should you enter the blast area with EMS personnel to "load and go" with the injured. Check others for signs of life, but leave the dead in place. Minimize the time in the blast zone. Remember Time, Distance, and Shielding.
- When feasible, stay upwind and uphill of the blast zone.
- *Consider secondary devices, armed attackers, or attackers carrying a PBIED.* Remember a suicide bomber may be among the injured.
- Stay out of mists, dusts, fogs, smoke, and dusts as they may be toxic.
- Use a radiation detection instrument or pager to determine if any radioactive materials are present.
- If persons are down and experiencing seizures, consider the explosion may be have been to disperse a nerve agent.
- Do not enter the area or have contact with anyone from the area without chemical protective clothing and at least an air-purifying respirator.
- Choose the best location for a command post outside the blast area and do not use it to avoid any preplaced IEDs. Look for out-of-place vehicles, objects, briefcase, bags, or other items. Have an officer run license plates of vehicles looking for switched tags, a stolen vehicle, or rental vehicle, which might contain a VBIED.
- If feasible, have victims remove their clothing as it may contain toxic materials.
- If clothing is removed, retain such for evidentiary purposes.
- The minimum perimeter should be at least 300 feet from the last site of broken glass (preferably 1,500 feet, but such may not be feasible).
- Establish staging areas outside this perimeter.

- If fires are present, they should be fought defensively and with remote appliances. If no fires are present and no one is in need of rescue, no one should enter the blast zone until the bomb squad has cleared it.
- If feasible and needed, have utilities secure water, natural gas, and power remotely.
- Remember, the blast area may contain additional IEDs, partially consumed explosives, toxic, or radioactive materials.
- When feasible, keep track of vehicles and persons leaving the blast site.
- The area should be treated as a crime scene. Fragments and evidence may be on building tops, in the clothing of victims, or on or within vehicles parked in the blast zone.
- Keep arterial routes open for ingress and egress of emergency vehicles.
- When feasible, locate and interview witnesses, collect video from adjacent businesses or other locations in the area.

Chapter 12

METHODS OF DAMAGE AND MEDICAL CONSIDERATIONS

Damage

Damage is caused by several methods. When low explosives such as gunpowder deflagrate (burn rapidly), they produce gases. The gases must be contained in a vessel (usually a metal pipe with end caps) and when the gas pressure overcomes the strength of the container, catastrophic failure occurs, propelling fragments from the pipe along with the escape of hot gases.

The primary cause of injury with low explosives is fragments unless the person is in contact with the device where burns from the hot gases may occur. The only work accomplished by low explosive is the production, albeit small amounts, of gases.

High explosives detonate and produce a transonic shock wave along with fragments and hot gases. The shock waves are shortwave phenomena, while fragments may be projected long distances. The shock wave damage is dependent upon the velocity of detonation (VOD). The higher the speed of VOD, the more brisant or shattering effect the shock wave has on the surrounding matrix. Lower VODs may range from 7,000 to 9,000 fps for dynamites, PETN around 25,000 fps, RDX around 26,000 fps, TATP around 15,000 fps, and TNT around 20,000 fps.

High explosives also produce gases, which produce most of the "work" performed by explosives. These gases "push" rather than shatter as does the shock wave, which precedes the gases. The gases are responsible for moving earth and rocks in blasting while the shock wave shatters the rock. By knowing the makeup of rock strata to be broken and using the appropriate VOD and gas-producing explosive, one can control the overbreak and size of rock fragments produced.

Explosives have individual group characteristics based upon their chemical composition. Overoxidized explosives, which carry their own oxidizers, produce gray to white smoke when detonated. Examples include ANFO, flash powder, most dynamites, and smokeless gunpowder. These explosives also tend to produce more gas than those explosives in other categories. Their energy is concentrated in the gases produced. This type of explosive is used commonly in mining applications because of the fewer toxic byproducts and work performed by the gases in breaking and moving material. Underoxidized explosives depend upon atmospheric oxygen and produce large amounts of nonoxidized products during the detonation. These are produced from the explosive, binders, fillers, container, and entrained organic material at the blast site. These products are carbon-based and produce a dark smoke and usually a large thermal event in the form of a fireball as the heat and unburned materials interact with atmospheric oxygen. In many instances, the energy liberated from the thermal event and fireball exceeds the energy from the shock wave and gases. Military explosives tend to be underoxidized and their thermal event used to destroy and set afire targets. Military explosives tend to have a high VOD as seen in the RDX and PETN explosives. Limited use explosives with even higher VOD can be seen in specialized applications such as nuclear weapons and special explosive devices.

One must remember that explosives are short-range weapons as their shock wave and gases rapidly degrade while interacting with the atmosphere and other objects. Intimate contact is needed for explosives to be effective. This is the reason an exclusion zone for vehicles is needed to prevent VBIEDs from coming in contact or penetrating a structure.

The shock wave is the compression of air in front of the transonic energy produced by the explosion. The shock wave travels near the speed of the VOD. This shock wave may produce overpressure in the several million pounds per square inch pressure. Shock waves can shatter concrete and masonry, deform metal while causing spalling, or shatter metal.

Shock waves degrade quickly but may operate in a quirky fashion as they are easily redirected and reflected. This may leave areas with minimal or no damage while concentrating damage in other areas. Some of the damage from shock waves is hypothesized to occur whenever the shock wave moves from one media to another. The interaction produces damage at and near the interface points cause spalling failures in solid brittle materials such as masonry, concrete, and metal. It appears when the shock wave changes velocity, energy is imparted to the media at this point.

Shock waves tend to follow terrain and wrap around objects. This may cause a vortex to form on the side of the object opposite to the shock wave. These vortices produce low pressure and cause damage to the object. This explains how objects have damage to surfaces facing away from the explosion. Damage may also occur at 90-degree angles from the shock wave and these are produced by the formation of vortices. Blast waves usually start in a spherical form but will rapidly conform to nearby objects, terrain, and are reflected or concentrated easily. The inverse square law applies as doubling the distance from the explosion will reduce the damage dramatically. This does not apply to projected fragments, which may travel long distances.

Shock waves also cause ground movement similar to earthquakes. Large VBIEDs may cause structural failure based upon ground movement. The shock wave movement through a structure also creates drag on adjacent walls, overhead, and floors. This drag may be sufficient to cause structural failure and was one of the primary causes of structural failure in the Oklahoma City bombing. The drag produced is toward the path of the shock wave.

To illustrate the short-range nature of explosives, the same amount of explosive placed one foot away from masonry wall may cause only surface damage. Placing the charge in contact with the masonry wall may cause severe damage. In dealing with explosives, distance is a lifesaver. The only long-range effect of explosives is the fragments projected.

Craters and ejected debris may be scaled to provide a rough estimate of the amount of explosives used (see Figures 112 & 113). For ANFO, dynamite, or TNT, the formulas below can be used with a crater or distant debris to estimate the amount of explosive used.

Most Distant Debris

ANFO or Dynamite WE = Explosive yield, r = distance most distant debris, K = scaling factor 91,000 M^3/kg, WE = $(R^3)/K$

Crater Sizing

Crater sizing in surface blast; d = crater diameter in meters W = explosion yield in kg of TNT; $D = 0.8(W)^3$

Formula for Fragmentation Distance of TNT

Fragmentation distance = 100 X cube root of weight of explosives in pounds is the fragmentation distance in meters

Crater Sizes (Surface Burst of TNT)

TNT lbs.	*Crater Width Meters*	*Crater Depth Meters*
1	0.074	0.28
5	0.076	0.30
10	1.47	0.46
20	2.01	0.74
30	2.29	0.94
40	2.33	1.07
50	2.95	1.21

Figure 112. Bomb technicians prepare to destroy approximately 50 pounds of old explosives.

Examples of Typical Explosives Effects

25 Pounds of High Explosives (Dynamite)

- Almost total damage to the floor of building
- Considerable damage two floors above and below the detonation point
- Expect fragment velocities of 6,000 fps
- Some fragments may penetrate masonry walls
- Extensive loss of glass in the structure likely

One Pound of High Explosive

- Thirty square foot room will be destroyed with masonry walls stopping fragments
- If the bomb is fragment enhanced, some fragment penetration of walls
- Glass shattered by shock wave

Figure 113. This is the resultant crater that was created from detonating the explosives shown in Figure 112.

Typical Pipe Bomb Using Smokeless Powder

- 900 square foot room will be damaged mainly by fragments
- Masonry walls will stop the pipe bomb fragments
- Damage will be limited to the room
- Glass likely shattered by fragments

Effects of blast overpressure from shock waves (psig–pounds per square inch above atmospheric pressure):

- 0.5 to 1.0 psig window glass is broken
- >1.0 psig people are taken off their feet and pushed down
- 1.0 to 2.0 psig structures with wood siding are damaged
- >2.0 to 3.0 psig cinder block walls with no reinforcement collapse
- >5.0 psig causes eardrum rupture
- >15 psig lung injuries begin
- >35 psig fatal injuries begin

- >50 psig LD50 (50% of those exposed are killed)
- >65 psig LD99 (99% of those exposed die)

Medical Considerations

- Mechanisms of injury
- Blast overpressure
- Thermal event
- Fragments penetrating victim
- Propelled objects
- Falling objects
- Structural collapse
- Falling or propelled glass
- Victim falling or propelled into objects
- Dust and dirt contaminating wounds
- Toxic combustion gases
- Toxic effect of explosives
- Other mechanisms of injury
- IED disperses a chemical agent
- IED disperses a biological agent
- IED disperses a radioactive material

Several mechanisms cause injury in the explosion environment. For those close to the explosion, blast overpressure in high explosives causes injury to hollow organs. These are called primary injuries. These are short-range injuries requiring a person be near the explosive device. In many instances, the fragmentation produces immediately fatal injuries. The shock wave injures the ears, sinuses, lungs, intestines. One of the suspected methods of damage to these organs is the change in velocity of the shock wave as it passes around and through the body's different media. Those liquid filled organs tend to have fewer injuries as velocity changes are not large while those hollow organs containing gases do experience substantial velocity changes with more energy imparted at the interfaces of the media.

Usually the first injury to appear of the hollow organs is eardrum rupture at about five psig. A person not hearing well, blood draining from the ears, and disorientation, may evidence this. Sinuses may be injured and this may be evidenced by blood draining from the nose and pain. A common finding in persons exposed to explosions is the loss of hearing. It does not by itself indicate a severe injury, but such should be considered. Those killed by overpressure usually have little external signs of injury. This is a form of blunt force trauma for which resuscitation is fruitless.

Lung injuries begin about 15 psig. Shortness of breath, pneumothorax (rupture of lung tissue releasing air into the chest), hemothorax (bleeding into the lungs), and tension pneumothorax (compression of the lungs by free air from ruptured lungs) are common injuries. Any of these, but particularly tension pneumothorax, can be fatal. Lung injuries are manifested close to the time of injury. Their mortality rate is high.

Rupture of the partially gas-filled intestines is possible. This type of injury is delayed in terms of symptoms appearing and may be manifested by abdominal pain. Other injuries may obscure this injury as their pain may override the abdominal pain, at least initially, of the gastrointestinal injury. The mortality rate of gastrointestinal injuries is high.

The thermal event of low explosives is small compared to that of high explosives, especially large amounts of the underoxidized or military explosives. Their fireball may project radiant heat for several hundred feet burning those exposed and setting objects on fire. Thermal burns, inhalation of toxic gases, and contact with the toxic unconsumed explosives may occur.

Secondary injuries occur from the fragments projected from the device, which cause penetration injuries. Any penetrating injuries from an explosion should be considered serious, as the external appearance may not show the depth and direction the fragment took. Serious wounds can exist below a small entry wound produced by fragments. In some instances, additional injuries are caused by objects propelled by the blast wave or gases produced, striking the person. Falling or propelled glass is another hazard as it may have a small surface injury overlying a serious internal injury. Glass tends to not x-ray well. Some attempts have been made to use glass as a fragment enhancer to defeat metal detectors. However, glass propelled from a device is a short-range fragment as the explosion tends to destroy the glass and with its low mass, it will not travel far.

Tertiary injuries come from the impact of the person with the ground, floor, or other upright surface. Blast waves will take people off their feet and their impact with the ground may produce blunt force

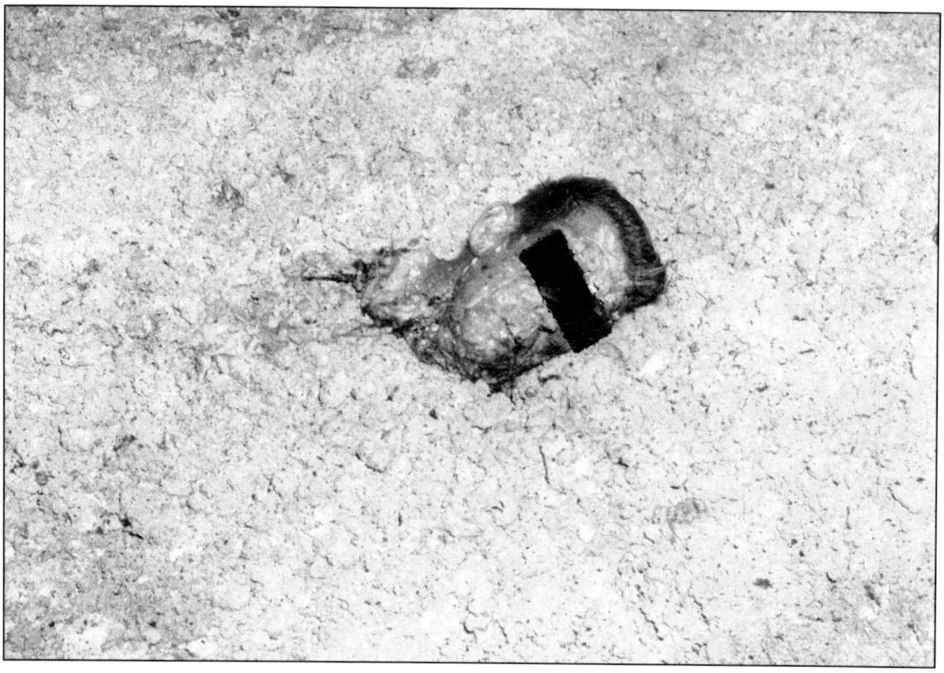

Figure 114. Careless use of explosives resulted in the victim's decapitation. (From Thomas Brodie, *Bombs and Bombings,* 3rd Ed. Courtesy of Charles C Thomas, Publisher, Ltd.)

injuries. Head injuries are common from this type event. Blast waves may cause a hydrostatic effect within the blood vessels injuring the brain.

Explosions produce a dusty and dirty environment. Any open wounds will likely be heavily contaminated. Decontamination by removal of clothing and rinsing with water is recommended when feasible. This will tend to remove any toxic materials such as partially consumed explosives or other agents. Pay special attention to the hair, which may collect and hold these contaminants.

Other injuries occur when structures collapse. Urban search and rescue along with heavy rescue units may be required to retrieve those entrapped by structural collapse. Crush injuries require special medical treatment, which includes large volumes of crystalloid IV fluids, sodium bicarbonate, pain medicines, and diuretics.

The clothing of victims may contain valuable evidence and should be retained. If wet, they should be allowed to air-dry. The bodies of the living and dead may contain valuable forensic evidence in the form of fragments.

Figure 115. This victim had been facing the explosive which was about four feet above the ground. The object of any bombing or accidental explosion investigation is to prevent more of the same from occurring. (From Thomas Brodie, *Bombs and Bombings*, 3rd Ed. Courtesy of Charles C Thomas, Publisher, Ltd.)

Figure 116. A 40mm grenade. These are spin-armed and very sensitive to movement once armed.

Situations in which explosive projectiles are lodged in the body of a living person are not unheard of and may require bomb squad support when surgery is conducted. A recent case in Alabama had a bomb technician assisting in the removal of a 40 mm grenade from the thigh of a person handling a found 40mm grenade. The propulsion charge ignited and propelled the grenade into the thigh of the person tampering with the grenade. The receiving hospital would not allow the patient entry into the emergency department until bomb technicians removed the grenade. This grenade is spin armed and once armed is very sensitive to impact (see Figures 114–116).

Chapter 13

RESPONSE TO EXPLOSIONS AND EVIDENCE CONSIDERATIONS

Many officers will respond to explosions during their career, but most will not be bombings. Some will be identified easily as not a bombing while others will not be as clearly determined. Officers should treat every explosion as a criminal or terrorist act until proven otherwise.

Following an explosion, accountability will be an issue. Occupants and bystanders will disperse. Many facilities will have a "distant" gather point at least 1,500 feet from the structure. Locating management or responsible persons will be a problem and is critical to determine if anyone may still be present in the structure or danger area. The first arriving officers may be overwhelmed with the injured or crowds. The first effort must be to communicate the severity and extent of the event. The additional resources needed should be requested.

The officer should approach upwind and uphill if feasible. Avoid physical contact with those leaving the site, especially if a number of people are down or are seizing. The officer should avoid smoke, dust, fumes, liquids, and mists. Remember, the bomb may be a dispersal device for chemical, radiological, or biological agents. Remember, Time, Distance, and Shielding! Minimize time in the danger area, stay as far away from the danger area is possible, and stay from under or near glass, keeping solid objects between you and the danger area. If you can see the suspicious device, you are within danger area. The blast site may contain additional devices, unconsumed explosives, contaminants, or booby traps. The blast area is a crime scene with valuable evidence.

Immediate Actions on Explosion Scene for Law Enforcement:

- Call for the needed resources
- Approach the scene cautiously and be observant for threats
- Law enforcement is the lead agency, coordinate with fire and EMS
- Use time, distance, and shielding
- Consider secondary devices and this is a lure to attack responders
- Consider this is an opening of multiple attacks
- Consider this is a distraction for a conventional criminal act (bank robbery)
- Stay out of the blast area unless rescue of injured is needed
- Don any needed PPE such as an APR or protective clothing
- Stay from under and away from glass
- Look for secondary devices, suspicious vehicles, or suspicious persons
- If multiple persons are down or seizing, do not enter the site
- Use a PA system to have the injured walk out to a segregation area
- Ask those down but not able to move who can hear you to raise their hand
- Use expedient methods to rescue those downed who cannot move
- Use "load and go" tactics to move the injured out of the blast area
- Remember, suicide bombers may be among the victims
- Leave the dead in place
- Keep ingress and egress routes to the site open
- Establish a guarded command post
- The next step is to **WAIT!**
- Conduct accountability of those within the facility
- Have the scene surveyed for radiation, chemical, or biological agents
- Use the time to establish an inner perimeter and obtain needed resources
- Establish a hard outer perimeter
- Stage the needed resources in presearched areas
- Have the bomb squad clear the site

Remember those exposed to an explosion may have hearing loss. These individuals may also have a decreased level of consciousness. Those dead on the scene cannot be resuscitated in most cases.

Glass breakage is a good method to determine the beginning of the danger area. Although fragments may be projected further than the shock wave, glass is broken with low overpressure. A good rule is, judge the distance from the epicenter of the explosion to the point glass breakage ends, multiply by 1.5 times to locate the outer perimeter. Inner perimeter location will vary but should be sited as far away as operationally feasible.

Immediate Actions on Explosion Scene for the Fire Service:

- Stage outside the blast area unless immediate firefighting is needed
- Coordinate with law enforcement and EMS
- Use time, distance, and shielding
- Call for the needed resources
- Approach cautiously observing for any suspicious persons, packages, vehicles
- Consider secondary devices and this is a lure to attack responders
- Consider this is an opening of multiple attacks
- Stay out of the blast area unless lifesaving efforts are needed
- Use defensive actions and fight fire with remote appliances from a distance
- Make certain all PPE is worn including SCBA
- Stay from under and away from glass
- Look for secondary devices, suspicious vehicles, or suspicious persons
- If multiple persons are down or seizing, do not enter the site
- Use a PA system to have the injured walk out to a segregation area
- Ask those down but not able to move who can hear you to raise their hand
- Use expedient methods to rescue those downed who cannot move
- Use "load and go" tactics to move the injured out of the blast area
- Remember, suicide bombers may be among the victims
- Leave the dead in place
- The next step is to **WAIT!**
- Conduct accountability of those within the facility
- Have the scene surveyed for radiation, chemical, or biological agents

- Use the time to acquire additional resources such as urban SAR or rescue
- Stage the needed resources in presearched areas
- Wait until the bomb squad clears the site to take any offensive actions

Immediate Actions on Explosion Scene for EMS:

- Stage outside the blast area unless immediate EMS assistance is needed
- Coordinate with law enforcement and fire service
- Use time, distance, and shielding
- Call for the needed resources
- Approach cautiously observing for any suspicious persons, packages, vehicles
- Consider secondary devices and this is a lure to attack responders
- Consider this is an opening of multiple attacks
- Stay out of the blast area unless lifesaving efforts are needed
- Make certain all PPE is worn such as chemical suits and APR
- Stay from under and away from glass
- Look for secondary devices
- Use a PA system to have the injured walk out to a segregation area
- Establish a triage area outside the blast zone in a presearched location
- If multiple persons are down or seizing, do not enter the site
- Ask those who can move to walk out of the blast zone
- Ask those down but not able to move who can hear you to raise their hand
- Use expedient methods to rescue those downed who cannot move
- Use "load and go" tactics to move the injured out of the blast area
- Leave the dead in place
- Remember the injured may be contaminated
- Survey victims for radiation, chemical, or biological agents
- Notify medical control and receiving hospitals of magnitude of event
- Stage the needed resources in presearched areas
- Remember, suicide bombers may be among the victims
- When feasible, remove and place the clothing of victims in a known location

- Remember clothing may contain contamination
- Remember valuable evidence may be within the clothing of victims
- Track where the injured are sent
- If feasible, stay 1,500 feet outside the blast area

Special Issues May Include:

- Crush injuries and those entrapped may need special treatment such as large amounts of IV fluids, pain medication, and a special treatment protocol may have to be used
- Minor surface wounds may overlie a deep penetrating fragment injury
- Those with hearing injury should be evaluated carefully
- Decreased level of consciousness is not uncommon but may be an indicator of traumatic brain injury
- Hollow organ damage, especially abdominal injuries, may not become symptomatic for several hours
- Open wounds will be contaminated with dirt, dust, and debris

Evidence may come in a variety of forms. Fragments may be obvious but some items may not be as obvious. The investigator can use assistance from the bomb squad and those investigators how have attended post blast investigation training.

Evidence and Evidence Collection

The primary consideration regarding evidence collection is officer safety.

Is The Scene Safe?

- Are any shooters or other threats present?
- Are any fires present extinguished?
- PPE will be need to include chemical protective clothing, puncture waterproof boots (if the scene is wet), nitrile gloves (consider puncture resistant over gloves), an APR or N99 respirator
- Confirm that radioactive materials, biological, or chemical agents are not present

- When surveying the scene and conducting a walk through, look for items out of the ordinary and consider what threats may be present
- Vehicles present should have their tags checked to confirm the tag matches the vehicle and the vehicle/tag is not stolen. If stolen this may conceal a VBIED.
- Rental vehicles should carry a high index of suspicion also
- Be alert for unconsumed explosives or IEDS
- Victims present could be booby trapped or be a suicide bomber

Where is the Scene?

The scene is considered:

- From the point that any windows are not broken to seat of blast plus one half the distance beyond the last broken window
- This will usually encompass most scenes
- Remember scenes tend to be circular with some fragments projected upward

What Does the Scene Encompass?

- Victims on scene
- Vehicles on scene
- Surrounding structures including rooftops
- Those victims or dead transported from the scene
- Secondary scenes may be at the hospital or morgue
- A tertiary scene may be the location the bomb was constructed
- PPE must be worn to protect officers and prevent cross contamination

Is Time a Factor in Examining the Scene?

Many items on the scene will be transient, so time is essential:

- Fiber and other trace evidence may be short-lived (less than six hours according to some authorities)
- Explosive vapors will be short-lived

Figure 117. An alarm clock prior to explosion.

- The seat of the explosion will likely be the area with the most damage and the explosion may cause a crater

Once the scene is identified and secured, one can begin the process of hazard elimination. One consideration is structural integrity of any involved structures. This may require and engineer to confirm such is safe to enter. Other issues may involve broken water or sewer lines flooding the scene. Firefighting efforts may have caused water to stand or washed away evidence. Identifying bomb components can be complicated. The rule is to collect any item that appears to be bomb related. Many extraneous items may be found in the explosive setting. Bomb components will usually show effects of the hot gases, flame impingement, high temperatures, and soot.

Bomb fragments are usually metal shards. If a low explosive is involved, the metal fragments will have rough brittle fracture edges and the pieces tend to be large. High explosives produce smaller fragments with the metal thinned less than 90-degree angles. The hot gases tend to erode the fragment edges in high explosives (see Figures 117–127).

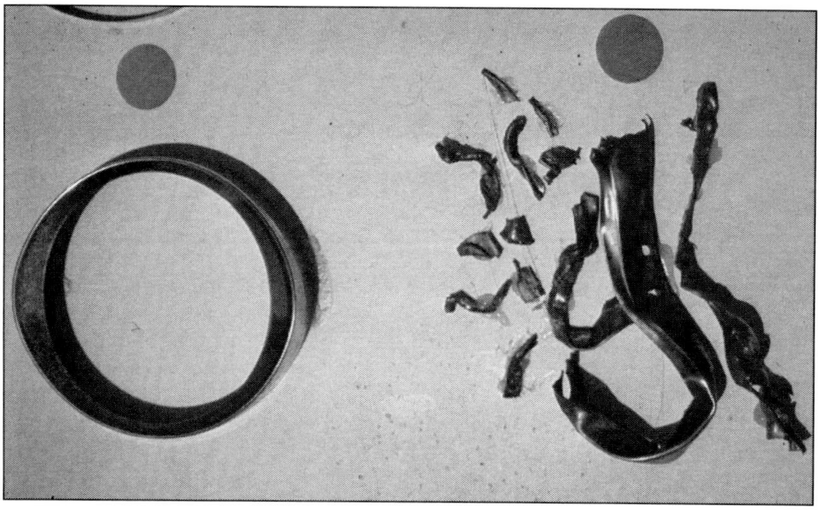

Figure 118. The alarm clock following an explosion.

Figure 119. Batteries following an explosion.

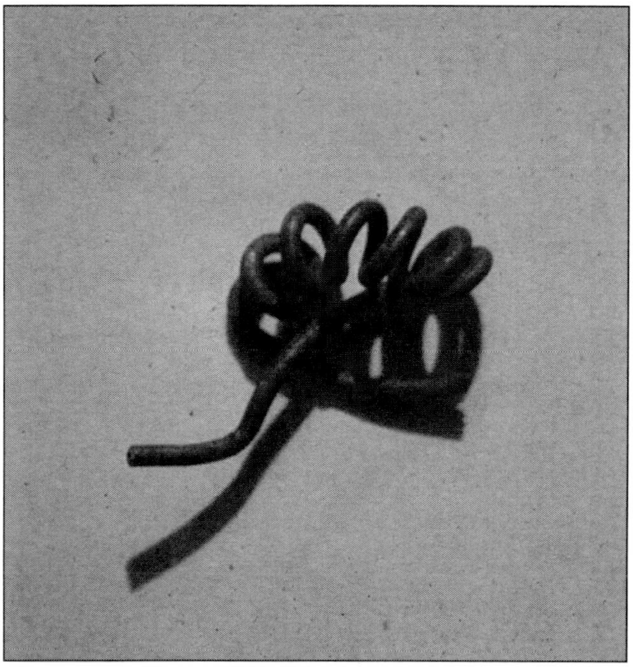

Figure 120. Remains of a clothespin spring following an explosion.

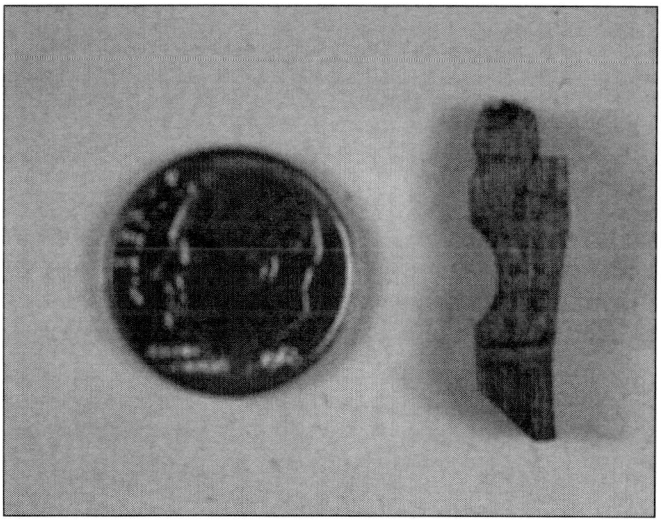

Figure 121. Remains of clothespin jaw following explosion.

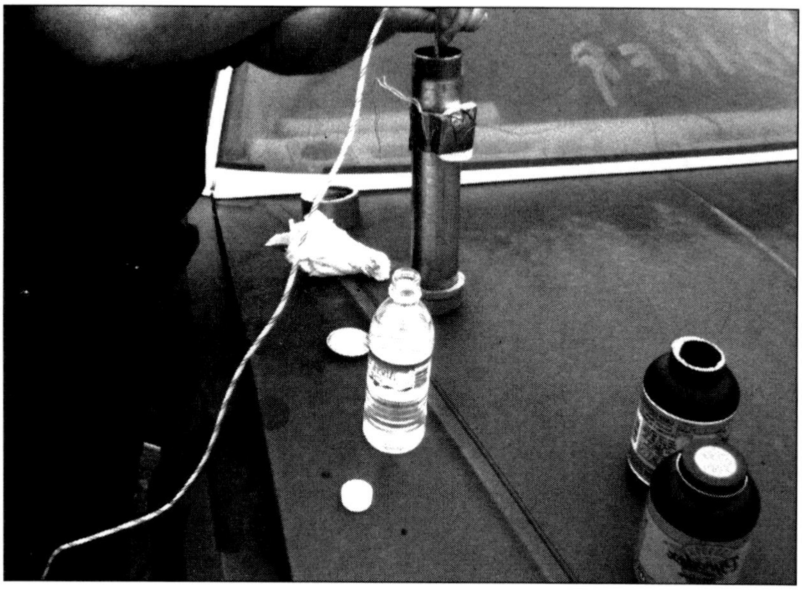

Figure 122. A bomb technician prepares a pipe bomb to destroy a vehicle during a post-blast evidence collection exercise. Several pounds of smokeless gunpowder were placed in a galvanized pipe. The pipe bomb was initiated using a blasting cap connected with detonation cord. (Photograph by Sergeant Gary Coleman.)

Figure 123. This is a vehicle to be destroyed by a pipe bomb preblast. The pipe bomb was placed under the driver's seat. (Photograph by Sergeant Gary Coleman.)

Figure 124. The pipe bomb explodes forcing the windshield out of the vehicle and projecting fragments through the roof. (Photograph by Sergeant Gary Coleman.)

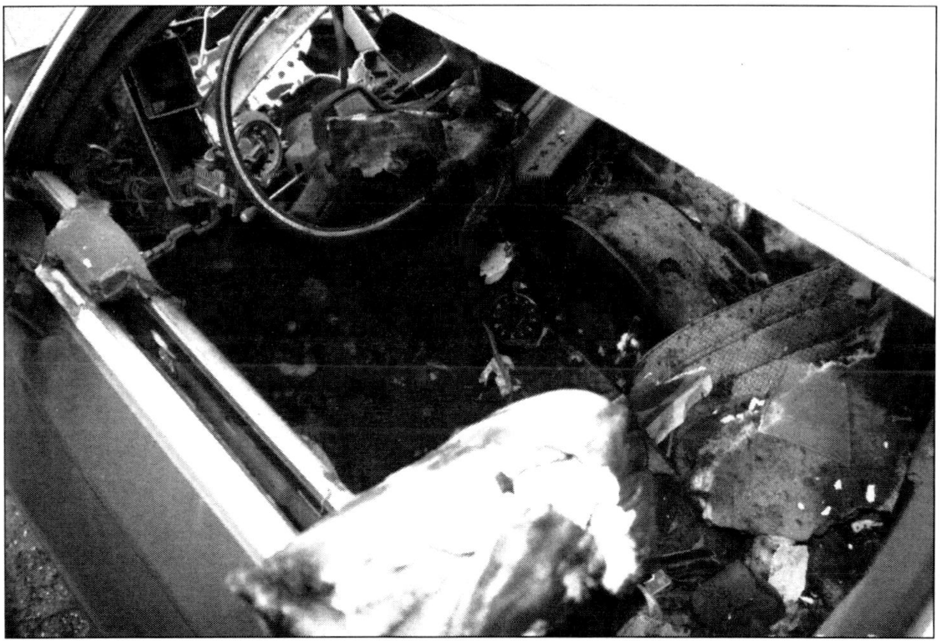

Figure 125. Blast effects from the pipe bomb. Note the floorboard is gone and a crater has been created beneath the car. (Photograph by Sergeant Gary Coleman.)

Figure 126. Note the bowed roof from the overpressure from the pipe bomb and the large fragment holes in the roof. (Photograph by Sergeant Gary Coleman.)

Figure 127. This is a bomb fragment roughly six inches in length ejected through the roof of the vehicle. This fragment was more than 300 feet from the explosion site.

The scene should be sketched and photographed. Measurements are important and the location of individual pieces of evidence is important. In some scenes, the trajectory of the fragments may be reconstructed. Scene mapping with laser systems are recommended. Collection of items should be conducted with care. Fragments can be sharp. Nitrile gloves are recommended. In handling fragments, use plastic tongs and avoid metal instruments to prevent damaging any tool marks.

When feasible, obtain polyester swabs of carbonaceous soot. These should be placed in an airtight clean metal can. Explosive vapors may be tested from the air space above the submitted material. Remember not to put wet items in sealed containers as this may cause bacterial and mold growth, which interferes with laboratory testing.

If in doubt as to the origin of a suspicious item found at an explosion site, collect the item. It can be examined later to determine its origin. Unconsumed explosives present a special hazard. First, the bomb squad should be called following isolation and evacuation. If the bomb squad deems the unconsumed explosive safe to sample, the protocol of the laboratory such is submitted to should be followed. The protocols vary and the investigator should be aware of the procedure. In most instances, no more than five grams is needed for laboratory analysis.

Investigators should wear one-time-use clean chemical protective clothing, shoe covers or clean chemical boots. Some investigators change gloves each time they have contact with any evidence to prevent cross-contamination. If the bomb squad is involved, bomb technicians will be familiar with cross-contamination issues and will use one-time-use equipment or equipment that has been rigorously cleaned to remove any explosive residue.

Exemplars of any materials used should be sent in separate packaging to the laboratory. Most investigators send clean sample containers and swabs, while others even send clean samples of gloves and PPE. The investigator should remember that body fluid contaminated items may present a blood-borne pathogen issue and wear appropriate PPE. Another consideration is that explosives are toxic as are their vapors. Respiratory protection should be considered in any explosion or post-blast setting. At least N99 respirators or air purifying respirators are recommended to protect from particulate hazards as a minimum. In rare cases, investigators have had to go to surgery or attend an autopsy to collect fragments. The chain of evidence in these cases will be important. These items will be a biohazard as they will be contaminated with body fluids. Incendiaries present a special issue. In many cases, the device itself may be consumed in the ensuing fire. However, in many circumstances the device may be reasonably intact.

Figure 128. This device was a plastic bottle filled with a flammable hydrocarbon. The device was ignited in a structure in an attempt to collect insurance. Note the footprints of firefighters who did not recognize the fire as an arson fire.

If a liquid accelerant is present, be familiar with the method the laboratory to be used requires it be collected, the volume needed, the type container needed, and delivery method. Some will require a capped glass vial contained within a plastic bottle filled with absorbent over packed in metal container for liquid samples. Some laboratories require hand transportation while others allow commercial services to be used. Molotov cocktails are a common incendiary device as are bottle bombs. Molotov cocktails can be in a variety of containers. Bottle bombs are usually plastic bottles, or less frequently glass bottles. In some instances, the plastic container may survive.

A factor in bottle bombs is they use corrosive materials in some configurations. These must be handled carefully to avoid chemical burns. In most of these devices latent fingerprints are destroyed by fire or the corrosive material (see Figures 128 & 129).

Some bomb components, once identified, might have the locale purchased identified. Several commercial databases and information

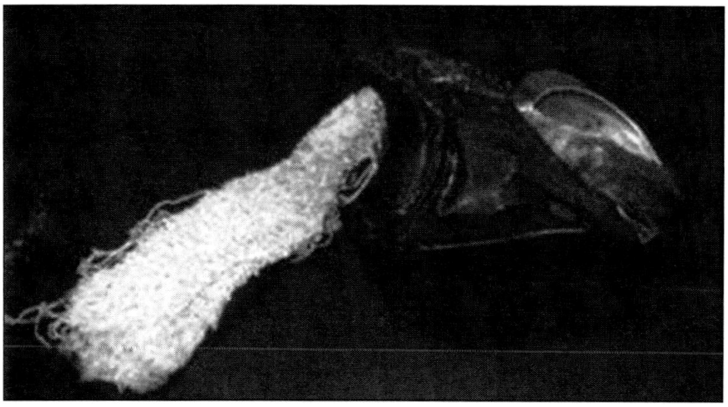

Figure 129. The adult who assembled this Molotov cocktail failed to realize metal cans are not frangible. He threw the incendiary device against his estranged spouse's residence several times with no results. In his frustration, he stomped the can, spraying burning gasoline over his clothing. When released from the burn center, he was arrested and charged with criminal possession of explosives and attempted arson of an occupied dwelling.

for law enforcement officers are available to assist in identification of bomb components. Some devices use components, which may be sold, locally at only a few locations. This may assist the investigator in suspect identification. Some bombers have been foolish and retained receipts for the purchase of bomb components. The UNABOMBER was aware of this process and hand built his devices avoiding commercial components. Another option is to use widely available components.

In some instances, the type device and "signature" construction of a bomber can be determined. The Bomb Data Center and Bomb Arson Tracking Center can determine if similar devices have been reported. Another asset is the FBI Special Agent Bomb Technician and Bureau of Alcohol, Tobacco, Firearms, and Explosives explosive specialist. The analysis of evidence can involve a multitude of tests. Vapors and residue may be analyzed for explosives using a gas chromatograph and atomic absorption spectrometer.

Fiber, DNA, and fingerprints may be obtained from some bomb components. This may be possible even postblast. Tool marks may be matched to tools used to construct the bomb should the bomb-making location be discovered. The laboratory may be able to determine the type components used as to brand and if commercial explosives are involved, the date and shift code for where and when the explosives were manufactured. Electrical tape and duct tape are fertile sources for fiber, DNA, and plastic fingerprints. They tend to survive explosions well. Labels and even envelopes or packages tend to survive explosions. They may provide the same types of evidence.

Chapter 14

PROTECTIVE MEASURES AND MITIGATION

Protective Measures

Perhaps the more concepts for protective measures are the time, distance, and shielding. This concept originates with radiation protection concept of **A**s **L**ow **A**s **R**easonably **A**chievable (ALARA). Time means spend the minimal amount of time in the danger area. Distance means stay as far away from the danger area or suspect item as feasible. Shielding means keep solid objects without glass between you and the suspect object. Some other rules:

- If you can see the suspect object, you are too close.
- If you can see the bomb technicians working, you are too close
- Following an explosion, if no lives are at risk, do not enter the site.
- Following an explosion, think about secondary devices.
- Avoid overhead glass.

Perhaps the best protection for static infrastructure such as structures is stand-off distance. This is especially true of structures where VBIEDs are a threat. The University of Wisconsin, Oklahoma City, and first World Trade Center bombings are examples of what VBIEDs can accomplish when stand-off distances are not used. Maintaining a stand-off distance for vehicles is critical.

The key is to prevent approach to the structures. A combination of active and passive features is needed to protect structures that include armed personnel and fixed or active features to prevent vehicle approach. Some facilities use checkpoints where vehicles are inspected before being allowed to approach the structure; use passive features such as bollards or glacis which prevent the approach of vehicles altogether.

The protection against PBIEDs is somewhat more complicated. Some entities use badging systems, which only allow personnel who have been vetted to enter the structure. However, the issue with checkpoints is that PBIEDs may be stopped, but the checkpoint, rich in human targets, becomes the target. If the PBIED is discovered, the suicide bomber detonates the device. One has to balance risks and realize that no matter what steps are taken, some degree of risk will remain.

Countersurveillance is the actions of personnel responsible for security but by every employee of a facility. They need to be alert for unusual events, persons out of the ordinary, persons asking unusual questions about the facility, or its personnel. Unusual attention to the facility or personnel, photography, or videography should be noted. Repeated visits or presence of persons with no apparent reason should be investigated. Mail should be screened with items not expected deemed suspect. The sender should be contacted to verify the package is legitimate. If such cannot be demonstrated, the package should be evaluated by the bomb squad. Some facilities x-ray incoming mail and packages and use a magnetometer to screen persons coming into the facility.

Employee protection should consist of training to recognize surveillance. Remote vehicle starting should be considered and vehicles should be parked in areas that are fenced and have surveillance systems. High-risk employees should have at least rudimentary courses on countersurveillance and a security assessment of their residence. High quality intrusion and panic alarms are recommended. Video surveillance is another item recommended to allow the resident to identify persons without allowing access. Bullet and intrusion resistant doors may be required along with reinforced and penetration resistant glazing.

High-risk facilities should also consider multiple layers of security including bullet and penetration resistant doors and glazing, along with intrusion and panic alarms. Valuable assets should not be located against exterior walls. Natural gas and electrical entry points should be protected against intrusion by vehicles and fenced. Video surveillance is recommended for these sensitive areas. Any hazardous materials or processes on site should be well protected and not located adjacent to exterior walls.

Bomb incident management training is essential for high-risk facilities. Bomb threat and reaction to suspicious persons and suspicious package training must be in place for every employee. Leaders should be schooled in the decision-making and risk management process in the event of a bomb threat, suspicious package, or person, and particularly regarding an explosion. An essential but forgotten group is the facility maintenance and janitorial staff. These individuals would likely note suspicious persons, suspicious vehicles, suspicious packages, plus they have access to the entire facility. These groups are also familiar with the facility and its support systems.

Exercises of a "realistic" nature should be conducted annually. The more self-sufficient a facility can be is important. A fire brigade and EMS squad can provide essential services until external assets can arrive. Employees must be familiar with predesignated assembly points including close and distant offsite locations. Employees should be educated to conduct a visual search of their workspaces prior to leaving, looking for suspicious persons, vehicles, or packages in the evacuation assembly areas. The leaders of a facility should be trained to evaluate threats and not evacuate blindly as a bomb threat may be a ruse to place employees in a position to be targeted by preplaced IEDs, shooters, or PBIEDs. Outside assistance can be obtained from the fire inspection services of many fire departments and some bomb squads will assist with bomb incident planning and bomb incident training. This advice is usually without cost and from individuals familiar with the regional threats and type IEDs in use.

Should an explosion occur, one must presume such is a bombing until proven otherwise. The results will likely include loss of power, fire, failure of sprinkler systems, structural collapse with trapped persons, and multiple injuries. Internal communication systems will likely not work. The environment will be dusty and dirty with open wounds contaminated with such. Even superficial appearing surface wounds may overlie a deep wound from fragments. Loss of hearing and confusion are common in the explosion setting. Glass shattering and falling produces many injuries.

Stand-off distance is important in VBIED attacks. Aesthetically pleasing concrete planters designed to resist vehicles can be used or conventional bollards can provide a gap between a potential VBIED and the structure. Connected parking decks, adjacent parking lots, or underground parking are a serious threat to the structure should a VBIED be used.

Positive identification of vetted staff and employees is crucial. At least a rudimentary background check should be performed on every employee. Positive identification is needed in the form an employee ID card. Some entities require the employee use an employee card with an RFID or magnetic stripe to gain access. Sensitive areas may require the same card be used but also couples a biometric measurement such as iris or thumbprint identification. Many of these systems are not expensive yet provide positive control of who accesses sensitive areas.

If a suspicious package is located, steps can be taken to mitigate potential blast damage. The important aspect to remember is taking this action places employees at risk and does reduce the fragment propagation and blast effects. These techniques also impair the operation of bomb technicians and their mitigation efforts.

Foam Blast Mitigation Systems

If the facility leader elects to expose employees to this danger, some of the techniques that can be used are the use of plastic garbage cans filled with water, use of commercial foam blast reduction systems, or the use of a ballistic blanket. These should be a last resort when no other assistance is readily available and the risk of exposing employees is acceptable. This should not be used when only property is at risk unless the damage to systems, which would endanger lives, are at risk and this is the only expedient method to mitigate the threat.

The mere placement of any item adjacent to or upon the suspicious object may cause the IED to function. Some agencies have used letter bomb carriers to isolate a suspected letter bomb and provide fragment protection while the facility is evacuated. The rationale is the item has already been handled and is not likely to be motion or movement sensitive. Ballistic blankets have also been used to provide some degree of fragment protection if evacuation must occur in proximity to a suspected IED.

Foam blast suppression systems are very effective. However, these systems may introduce weight onto the device and cause the device to detonate. The water-based foam may also cause a short circuit in elec-

Figure 130. This vehicle contains a pipe bomb that is located underneath the driver's seat. The bomb is filled with smokeless gunpowder and is ready to be initiated by blasting cap and detonation cord. This is a similar device and similar vehicle as seen in Figures 124 and 125.

Figure 131. The aftereffects of the explosion of the vehicle pipe bomb. Note that little foam displacement occurred and no fragments exited the vehicle as compared to Figure 126 where fragments penetrated the vehicle body and were found some 300 feet away.

Figure 132. An interior shot of the pipe bomb explosion within the vehicle. Note minimal damage to the driver's seat as compared with Figure 125.

tronic components of an IED should the foam unintentionally be exposed to them. This may cause a detonation. Alcohol-resistant class A foams are recommended as published research conducted by the author has demonstrated successful blast suppression and diminished fragment propagation (see Figures 130–136).

Figure 133. Note the large part of the pipe bomb still intact. Fragments of the pipe bomb used in the nonfoam car explosion were found 300 feet away, whereas this fragment remained within the vehicle.

Figure 134. A large portion of the pipe bomb was recovered from within the vehicle.

Figure 135. A common failure point on metallic pipe bombs is the end cap. Note the gases forced the endcap upper cap to fail. Note the endcap is relatively intact.

Figure 136. Other large pieces of the pipe bomb are recovered from within the vehicle.

Checklist for Facility Protection:

- Is the facility equipped to operate without power, computer support, water, sewer, and communications from external sources?
- Is an effective notification system and backup plan in place for notification of employees?
- Is the evacuation and accountability plan workable?
- How is communication achieved during and after evacuation?
- Does an onsite plan include basic EMS and firefighting procedures and equipment?
- Does well-defined "chain of command" exist when leaders are not present?
- Are sensitive and external areas well protected?
- Are any hazardous materials present protected?
- Is the facility leadership concerned about safety?

Chapter 15

CLANDESTINE DRUG LABORATORIES

The rise in the number of clandestine drug laboratories and their multiple uses make this a necessary component of responding to bomb-related incidents. Laboratories are used for multiple reasons, including production of controlled substances such as methamphetamine and its analogs; production of THC analogs and other controlled substances; production of homemade explosives (HME); production of biological weapons, poisons, or toxins; and the production of chemical weapons. Rarely, one might encounter a laboratory such was seen in the juvenile who produced a nuclear reactor-type device described earlier in the text.

The responder now needs to differentiate the types of laboratories present. The more common laboratory encountered will be a methamphetamine or laboratory producing some other type of controlled substance. Methamphetamine laboratory recognition is discussed in the next section. The more likely used process to manufacture methamphetamine is commonly known as the "shake and bake" method. This uses a plastic bottle with the methamphetamine manufacturing reagents added together. The signature for this laboratory will be the common methamphetamine components and a plastic bottle bubbling. If the plastic bottle is capped, the pressure may cause a catastrophic structural failure spraying a flammable volatile organic compound into the immediate area. A fuel air explosion will likely follow this container failure with an even larger and more intense thermal event setting fire to surrounding area. For this reason, when an active "shake and bake" cook with a capped container is discovered, iso-

Figure 137. An aerial shot of a mobile home that exploded from the operation of a red phosphorus methamphetamine laboratory. Note the burned trees and fire following the explosion. This explosion awakened the author who lives some four miles away. (Photograph by retired bomb technician Corporal John Skipper.)

late the container without moving the container especially if the container is bulging, evacuate the area, with fire department assets staged nearby. Only a person in a fire and chemical-resistant suit, wearing a self-contained breathing apparatus, should approach such a device (see Figures 137 & 138).

If the container is not capped, the area should be isolated and evacuated to at least 100 feet. Fire department assets should be staged nearby. The hazard will be lessened once the reaction is complete. This reaction evolves volatile ammonia compounds such that areas downwind may have to have a larger evacuation area. Thermal imaging and an infrared thermometer may be used to determine when the container is sufficiently cool to approach. Even then, fire resistant chemical clothing and wearing a self-contained breathing apparatus are recommended.

Figure 138. Photograph of the remains of a mobile home involved in an explosion. A red phosphorus laboratory was in operation when the explosion occurred. A microwave was in use while vapors from acetone were present causing the explosion. One laboratory operator was killed and the second burned. (Photograph by retired bomb technician Corporal John Skipper.)

The signature of a "shake and bake" laboratory will be a reaction vessel, typically a plastic soft drink bottle filled with liquid and bubbling while generating heat. Floating bits of gray colored material may be floating and circulating. This will be lithium metal. Sparks within the container may be observed when the lithium reacts with the reagents. These will not propagate into an explosion simply because the mixture is too rich with fuel to support the explosion. A common component found in almost every methamphetamine laboratory will be an acid generator. This is usually a plastic gasoline container with hoses protruding. This will be used to mix salt and sulfuric acid, which produces hydrogen chloride gas. The gas is used to bubble the gas through methamphetamine "oil" to convert it to the hydrogen chloride form, which precipitates as crystals. The crystals are filtered to obtain the final product (see Figures 139 & 140).

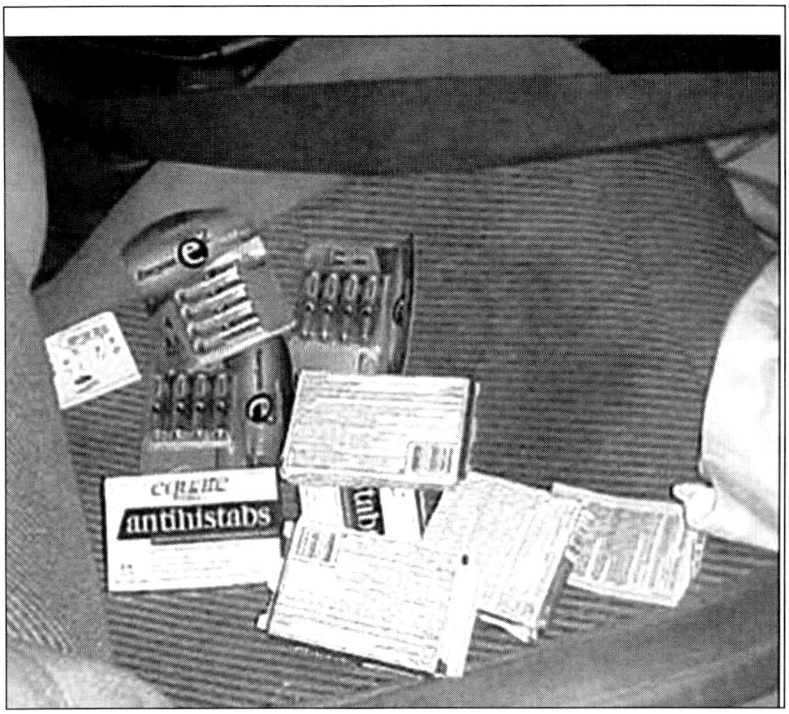

Figure 139. Lithium batteries and pseudoephedrine are two common components of methamphetamine laboratories found in "shake and bake" laboratories.

Items likely to be found in a "shake and bake" methamphetamine laboratory include the following:

- Ephedrine or pseudoephedrine, source cold and allergy tablets
- Lithium, source lithium batteries
- Starter fluid (petroleum ether) or ethyl ether, source starter fluid
- Ammonium nitrate source cold packs or fertilizer, ammonium sulfate fertilizer
- Sodium hydroxide, source Red Devil Lye
- Sulfuric acid, source batteries or sulfuric acid drain cleaner
- Table salt
- Acid generator

Other methods of production of methamphetamine include the Red Phosphorus or Red "P" method. This involves a more labor-intensive method. The danger in this laboratory is the production of phosphine gas, which can be fatal if inhaled. Items, which tend to differentiate this

Figure 140. A typical acid generator is shown, which is another item most often found in most methamphetamine laboratories.

laboratory from other methamphetamine laboratories include red phosphorus from match striker strips and a heat source. This method is no longer the preferred method but may still be encountered (see Figures 141–144). Items to look for include:

- Ephedrine or pseudoephedrine, from allergy or cold tablets
- Red phosphorus, source match striker strips (match boxes with the striker strips gone, ignition strips from road flares
- Organic solvents such as acetone or methyl ethyl ketone (MEK), source fingernail polish remover, lighter fluid, toluene, HEET gas additive, white gasoline, camp fuel
- Iodine, can be derived from iodine tincture by mixing with hydrogen peroxide
- Sodium hydroxide, source Red Devil Lye
- Sulfuric source, batteries or sulfuric acid drain cleaner
- Rarely hypophosphorus acid, explosion hazard
- Table salt
- Heat source such as a coffee maker or microwave oven

Another technique for manufacturing methamphetamine is known as the Birch or "Nazi" method. This involves using anhydrous ammonia, which is very dangerous. The presence of anhydrous ammonia is the signature to this type laboratory. Although not a common tech-

Figure 141. This commercial grade iodine is often found in red phosphorus laboratories.

Figure 142. Match striker strips have been removed. This is common in a red phosphorus laboratory but could be a sign of an explosive laboratory producing Armstrong's reagent using red phosphorus.

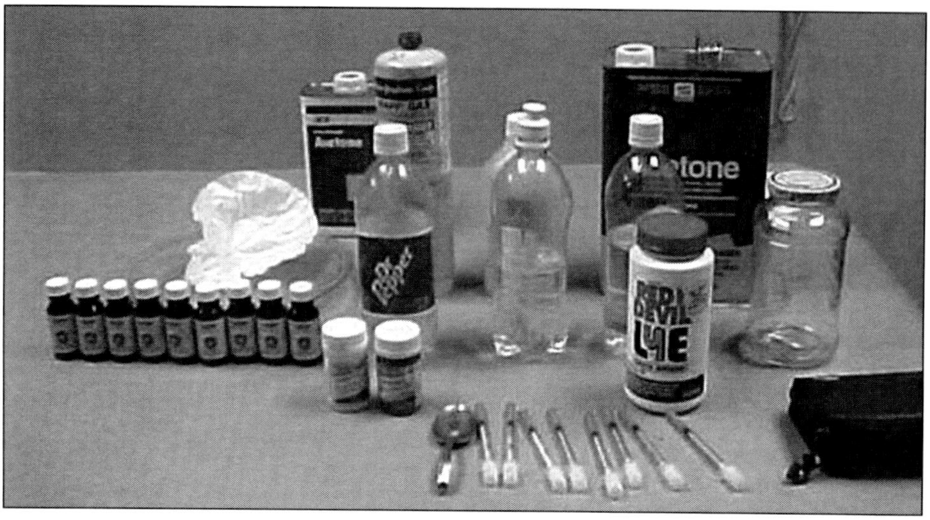

Figure 143. A simple red phosphorus laboratory using hydrogen peroxide to remove the iodine from iodine tincture. The operator was also using methamphetamine via the subcutaneous route as evidenced by the syringes.

Figure 144. What appears to be trash is a red phosphorus laboratory which has been discarded. Note the empty pseudoephedrine containers, empty match books with striker strips removed, hydrogen peroxide, lighter fluid as a solvent, and pile binders in the plastic bottle.

nique, it does produce high-grade methamphetamine. This lab produces an ammonia odor. Look for the following:

- Ephedrine or pseudoephedrine, source cold and allergy tablets
- Anhydrous ammonia, source stolen from agricultural supply facilities or farms, likely contained in an LPG cylinder. The anhydrous ammonia will discolor the spark resistant valve to a bluish color; however, it may be contained in a thermos or cooler.
- Lithium, source lithium batteries
- Organic solvents such as acetone or methyl ethyl ketone (MEK), source fingernail polish remover, lighter fluid, HEET gas additive, lighter fluid, white gasoline, camp fuel
- Sodium hydroxide, source Red Devil Lye
- Sulfuric acid, source battery acid or sulfuric acid drain cleaner

Items common to most methamphetamine laboratories include the following:

- Blenders or other grinding apparatus to grind the ephedrine or pseudoephedrine tablets
- Acid generators which are usually plastic gasoline containers or similar plastic containers with tubing
- Coffee filters to strain the pill binders

Unusual methamphetamine laboratories, which are not likely to be encountered, include several listed below. The common items found in methamphetamine laboratories will be present but additional items will be found also. The distinguishing configurations or materials are noted (see Figures 145–153).

Methylenedioxyamphetamine (MDA), Ecstasy Laboratory

- Isosafrole
- Formic acid
- Hydrogen peroxide
- Benzene (carcinogen)
- Methanol
- Ammonium formate

MDA Analogs

- Piperonal
- Nitroethane
- Ammonium acetate
- Lithium aluminum hydride (explodes on contact with water)

MDMA Laboratory

- Piperonylacetone
- Methylamine
- Sodium cyanoborohydride

MDA Using No Ephedrine or Pseudoephedrine

- Helional or floramelon, over-the-counter chemical used in perfumes
- Sodium hypochlorite, bleach
- Nickel acetate
- ETOH, drinking alcohol
- Acetic acid, vinegar
- Nitromethane, racing fuel
- Sodium hydroxide, Red Devil Lye
- All noncontrolled over-the-counter chemicals
- Converts the helional or floramelon to MDA
- "Shaker" or Thionyl chloride laboratory
- Paint shaker and Thionyl chloride are the signature of this laboratory
- Unusual solvents such as chloroform (carcinogenic)
- Palladium black may be used as a catalyst
- Hydrogen gas in a red cylinder may be present

Phenyl-2-Propanone (P2P, Propadope, Biker Lab)

- Methylamine an mercuric chloride (very toxic and corrosive)
- Aluminum foil

Phenylacetic Laboratory

- Phenylacetic acid (corrosive, easily absorbed through skin)
- Acetic anhydride (very corrosive, reacts violently with water)

Phencyclidine (PCP) Laboratory (Level A PPE recommended)

- Piperidine
- Cyanide (very toxic)
- Sodium bisulfate
- Magnesium metal (flammable, cannot put out with water)
- Petroleum ether

Gama Hydroxy Butyrate (GHB) Laboratories

- Gama butyrol lactone
- Sodium hydroxide (lye)
- Volcano method
- Mixes components of Red P laboratory together
- Heated with torch for a few seconds
- Uses chloroform (carcinogen) to wash out methamphetamine

LSD Laboratory

- Lysergic acid
- Acetonitrile
- Trifluroacetic anhydride (reacts violently to water)
- Diethylamine
- Benzene (carcinogen)

Methathinone (Cat) Laboratory

- Sodium dichromate
- Chromium trioxide (Toxic)

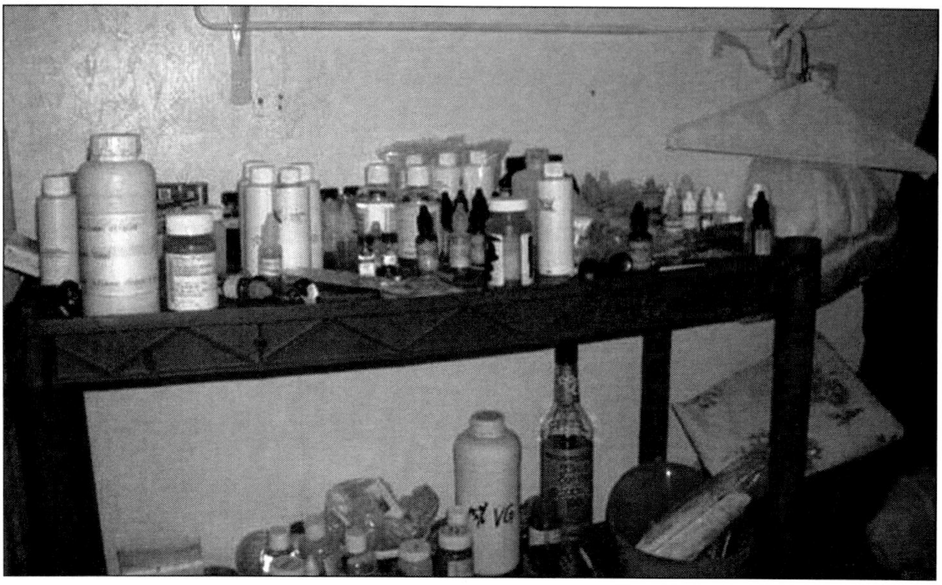

Figure 145. A laboratory producing synthetic marijuana.

Figure 146. The burning of precursor and laboratory material is common. The remains of containers of precursors, batteries, and the like may provide an indication as to what type laboratory is present.

Figure 147. Officers executed a search warrant and had information that an underground laboratory was present. Officers became suspicious of this doghouse when they discovered that it was too small to accommodate the dog present.

Figure 148. Officers become very suspicious when wiring was noted within the doghouse.

Figure 149. The doghouse was removed remotely and an underground laboratory was revealed.

Figure 150. Note the seals around the door and poorly constructed wooden ladder. Officers tested the atmosphere remotely and used a camera system to assess the laboratory remotely. This is a confined space, which required a backup rescue team with appropriate equipment.

Figure 151. The author and bomb technician Corporal Chris Vida suit up to make an entry into the underground laboratory. Out of frame is the standby rescue team and rescue equipment. The atmospheric tests were normal and less than four inches of earth covered the roof of the laboratory, which was well constructed and used a modified concrete septic tank as its structure. The concern was from mold since the laboratory was wet and officers wore SCBAs and chemical suits.

Figure 152. The interior of the laboratory, while crude, was functional and had power.

Figure 153. What concerned officers was the sophisticated surveillance system presence including passive infrared sensors in several locations surrounding the laboratory. The dog house entrance can be seen in the lower left of the photograph.

Butane Hash Oil (BHO)

This is a process by which the THC is "washed" from marijuana using liquid butane. Marijuana is placed in a tube into which butane in liquid form is introduced. The butane "washes" the THC from the marijuana in the form of a gooey-type liquid resembling honey. This process can be accomplished using carbon dioxide or an organic solvent such as acetone. However, the primary method is with butane. Butane is flammable and in an enclosed space explosive. The hallmark of this type laboratory is explosions. Traditionally hash is made from the female cannabis plant trichrome (see Figures 154–156).

The THC content can be above 85% in these products. One ounce of high-grade marijuana will make roughly six grams of hash oil. The price varies upwards from $45 per gram. BHO may be incorporated into candy. Some of the candy may have as much as 50 mg of THC. One issue is the extractor tubes resemble pipe bombs. Further, when

Figure 154. An odiferous compound is usually contained in this bottle. This was intended to discourage officers from opening the bottle. The suspect concealed BHO within the bottle.

Figure 155. Note the honey-like appearance of the butane hash oil.

Figure 156. Some individuals may dip cigarettes or cigars in BHO or synthetic marijuana compounds contained in liquids. Both of these items had been dipped in both BHO and a synthetic marijuana compound dissolved.

these operations cause a fire and explosion, responders may find the extraction tubes, which come in glass, metal, and plastic with end caps or end caps with holes in them. They may believe these are IEDs especially if an explosion is involved. In many instances, these operations are discovered when the butane used is ignited. Explosions within refrigerators where the butane may be stored or where extractor tubes may be placed have been reported. Another serious hazard is the BHO may be part of the toxic gases emitted in the fire. Some agencies have reported smoke and fumes containing very high levels of THC.

Anhydrous Ammonia

Anhydrous ammonia, rapidly changed to ammonium hydroxide, is evolved during the "shake and bake" method and evacuations may be required when such occurs. Usually the area evacuated will be small. However, the more serious hazards come from larger operations using the Birch or Nazi method of manufacturing methamphetamine where more substantial amounts of anhydrous ammonia are used. Although these methods have been supplanted by the shake and bake meth-

Figure 157. Note the bluish discoloration, which indicates that anhydrous ammonia has likely been stored in this cylinder.

od, they are still in use. Anhydrous ammonia is typically stolen from a farm operation from tanks, which come in a variety of sizes. Fixed facilities such as fertilizer manufacturing facilities may have tanks of anhydrous ammonia. The author was present when a large fixed tank of anhydrous ammonia was released by individuals attempting to steal the anhydrous ammonia for manufacturing methamphetamine. The release resulted in an evacuation of a several square mile area in a metropolitan city. When stolen, anhydrous ammonia may be stored in a thermos, cooler, or more commonly an LPG cylinder. If stored in an LPG cylinder, the nonsparking brass valve will turn a bluish color. This corrosion will eventually cause a valve failure. Individuals transporting anhydrous ammonia in a enclosed vehicle have been killed when the anhydrous ammonia was unintentionally released. Even a few breaths in an anhydrous ammonia atmosphere may produce sufficient to damage to the lungs to cause death or permanent disability. Several law enforcement officers have been severely injured inhaling anhydrous ammonia (see Figure 157).

If the theft of anhydrous ammonia occurs, one consideration should be its use in the manufacturing of methamphetamine, but consider it might be used as a toxic industrial chemical and as an improvised choking agent. Opening containers or cylinders of anhydrous ammonia upwind of an outdoor gathering would induce panic and trampling. If used indoors, it might cause sufficient levels to seriously injure eyes and lungs or cause death. On contact with water, anhydrous ammonia becomes ammonium hydroxide and is very basic and corrosive. Thus, breathing creates ammonium hydroxide in the lungs and contact with the eyes or damp skin creates the same effect. This corrosive effect injures the tissue. Reports of persons with anhydrous ammonia injuries should pique the attention of law enforcement personnel as this might be an indicator of methamphetamine manufacture or acquisition of the agent for WMD purposes. One should remember anhydrous ammonia is a cryogenic liquid and will produce freezing injuries on contact.

The United States Department of Transportation classifies anhydrous ammonia as a nonflammable gas in Class 2.2. This means the placard will have a green background and the four-digit code "1005" may be present on the tank or container may have Anhydrous Ammonia written on it. Even though labeled as a nonflammable gas, under certain conditions, anhydrous ammonia will ignite and explode. The conditions conducive to an anhydrous ammonia explosion include confinement and a rich environment. The LEL of anhydrous ammonia is around 15,000 ppm according to the CDC with its IDLH at 300 ppm. Most individuals can smell ammonia at two to five ppm. The LEL is about 15% and the UEL is roughly 28% according to NIOSH. This requires a rich environment with a very hot ignition temperature of around 1200° F. When released under pressure, anhydrous ammonia has a temperature of -28° F, its boiling point. Anhydrous ammonia has a liquid to vapor ratio of 850 meaning one gallon of anhydrous ammonia will produce 850, gallons of vapor. Anhydrous ammonia has a higher vapor density than air and will settle to the ground when released, but one good factor regarding anhydrous ammonia is the sharp warning odor.

If the odor of ammonia is encountered, officers should evacuate upwind and uphill. The potential for serious lung and eye injuries is high. Those exposed to anhydrous ammonia should have their clothing removed and eyes irrigated for at least 20 minutes. Wet decontamination is essential in anhydrous ammonia exposures. Prompt medical attention in a hospital setting is needed.

Clandestine Drug Laboratory Hazards

The hazards may be physical, biological, or chemical. Perhaps the more serious hazard is the operators of the laboratory who might be armed and likely under the influence of methamphetamine. Physical hazards include substandard structures, explosive atmospheres, improper wiring, wet floors, and oxygen deficient atmospheres. Chemical hazards may include leaking chemical and gas containers, incompatible chemicals stored together, improper mixing of chemicals, carcinogens, and other toxic chemicals present. Biological hazards may include mold and other fungus growths in the laboratory setting, especially if the laboratory is poorly ventilated and situated underground.

Booby traps are another hazard. These come in a variety of forms including mechanical, explosive, and chemical. The booby trap could be a deadfall with nails in a board in the bottom of a concealed pit or fish hooks on monofilament line. Electronic alarms may be present to alert operators of intrusion, or explosive devices designed to injure or kill intruders. Well-sealed or below ground level laboratories may be oxygen deficient or contain toxic gases. Air sampling and the wearing of at least Level B PPE are recommended when feasible.

Heat stress from working in PPE is another threat. Officers should operate in pairs using the "buddy system" to check on the partner. If signs of heat stress such as nausea, dizziness, rapid pulse, or weakness are experienced, the partner should evacuate the affected officer.

INDEX

C

D

Q

R

S

T

ABOUT THE AUTHOR

Jim Smith has more than 40 years law enforcement and public safety experience. He worked his way from police officer to police chief and finally public safety director. Smith remains an active police officer, firefighter, and paramedic. He is a certified police chief, APOSTC certified law enforcement executive, Nationally Registered Paramedic, Advanced Certified Law Enforcement Planner, Pro-Board certified Fire Instructor, and is a Professional Member of the American Society of Safety Engineers. He also serves as a fusion center liaison officer. He is a long-time member of the International Association of Bomb Technicians and Investigators and International Society of Explosives Engineers.

Colonel Smith has a master's of science in safety from the University of Southern California, along with a bachelor of science in biology from Troy University, and an associate in science in emergency medical technology. He serves as the public safety director for the Town of Cottonwood and as public safety coordinator for Troy University. Smith organized and supervised the Dothan Police Bomb Squad, Clandestine Laboratory Entry and Assessment Team, and the WMD Response Team for 10 years. This bomb squad continues to serve a 10-county region and is FBI certified.

Smith has responded to more than 100 bombings and incendiary events. He recently prevented an attempt to bomb a school. He executed a search warrant, which seized unassembled bomb components from a young offender alleged to have plotted to manufacture a bomb and deploy the IED at a local high school. This is the second incident in which unassembled IEDs, allegedly slated for deployment at a high school, was disrupted by Colonel Smith. He has dealt with numerous clandestine drug laboratories and marijuana fields, which were booby trapped. He has also dealt with several homemade explosives manufacturing incidents. Smith developed and teaches a certification program for clandestine laboratory safety and assessment and also developed and teaches a CBRNE certification program for a university and state emergency management agency.

Smith teaches as an adjunct instructor in criminal justice, terrorism, and homeland security for Troy University and University of Phoenix. He has published five textbooks addressing homeland security, weapons of mass destruction, biological and chemical warfare, radiological incidents, bomb incident management, tactical medicine, and crisis management. Smith is a Diplomat of the American Board for

Certification in Homeland Security and sits on the board of directors for the American Board of Law Enforcement Experts. He serves as a peer reviewer for *Inside Homeland Security* and for several criminal justice publishers. Smith has served as a jail commander, 911 district coordinator, 911 center director, and health physics technician, radiochemist, and radiological defense officer. He is a graduate of the US Army Chemical School, numerous homeland security programs, and is an instructor for several homeland security-related programs. Smith has developed several programs including university level courses relating to criminal justice, terrorism, emergency management, and homeland security along with clandestine laboratory safety and assessment.

Colonel Smith has published more than 50 papers in professional journals including research into methods for blast and fragment mitigation, which led to a patent for the process. He recently provided expert testimony, which led to the issuance of a patent for a proprietary explosive device.